EXPERIENCING GEOMETRY

EUCLIDEAN AND NON-EUCLIDEAN

WITH HISTORY

EXPERIENCING GEOMETRY

EUCLIDEAN AND NON-EUCLIDEAN WITH HISTORY

THIRD EDITION

David W. Henderson
Daina Taimiņa
Cornell University, Ithaca, New York

Upper Saddle River, New Jersey 07458

Library of Congress Cataloging-in-Publication Data

Henderson, David W. (David Wilson)
    Experiencing geometry : Euclidean and non-Euclidean with history / David W.
      Henderson, Daina Taimina.
        p. cm.
    Includes bibliographical references and index.
    ISBN 0-13-143748-8
    1. geometry. I. Taimina, Daina. II. Title.

    QA453 .H497 2005
    516-dc22   2004053105

Editor-in-Chief: Sally Yagan
Executive Acquisitions Editor: George Lobell
Vice President/Director of Production and Manufacturing: David W. Riccardi
Executive Managing Editor: Kathleen Schiaparelli
Senior Managing Editor: Linda Mihatov Behrens
Production Editor: Rob Walters, Prepress Management, Inc.
Assistant Manufacturing Manager/Buyer: Michael Bell
Manufacturing Manager: Trudy Pisciotti
Marketing Manager: Halee Dinsey
Marketing Assistant: Rachel Beckman
Art Director: Jayne Conte
Cover Designers: the authors

©2005, 2001, 1996 Pearson Education, Inc.
Pearson Prentice Hall,
Pearson Education, Inc.
Upper Saddle River, New Jersey 07458

Portions of this material are based upon work supported by the National Science Foundation under Grants No. USE-9155873 and DUE-0226238 and by D. D. Eisenhower Title IIA grants administered by the New York State Department of Education. Any opinions, findings, and conclusions or recommendations expressed in this material are those of the authors and do not necessarily reflect the views of the National Science Foundation or the New York State Department of Education.

Printed in the United States of America

10 9 8 7 6 5 4 3 2

ISBN   0-13-143748-8

Pearson Education LTD., *London*
Pearson Education Australia PTY, Limited, *Sydney*
Pearson Education, Singapore, Pte. Ltd.
Pearson Education North Asia Ltd., *Hong Kong*
Pearson Education Canada, Ltd., *Toronto*
Pearson Educaciûn de Mexico S.A. de C.V.
Pearson Education — Japan, *Tokyo*
Pearson Education Malaysia, Pte. Ltd.

To all the students who have studied geometry with us
(you have taught us much about geometry)

and

To our children and grandchildren:
Keith, Becky, Lelde, Linda, Lisa, Abigail, Erin, and Liam

PHOTO AND FIGURE CREDITS

Front Cover: designed by the authors and produced from the authors' photographs, digitally modified by the authors.

Page xxx: photos by Daina Taimiṇa.

Figure 0.1: photo by Daina Taimiṇa.

Figure 0.2: digital produced image by David Henderson based on painting in the Sistine Chapel, Vatican.

Figures 0.3, 2.1, and 14.5: digitally modified by David W. Henderson based on photos by Daina Taimiṇa.

Figure 0.4: image courtesy of NASA at http://map.gsfc.nasa.gov/m_mm.html

Figures 0.5, 21.10: photos by David W. Henderson.

Figure 0.6: photos by David W. Henderson

Figures 0.7, 1.4, 1.5, 21.5: photos by authors of models in Cornell University's Reuleaux Kinematic Model Collection.

Figures 0.8, 21.7b: photo by Manfred Mornhinweg (La Serena, Chile). Used by permission.

Figure 1.3: from Villard, de Honnecourt, 13th Century. *Album de Villard de Honnecourt, architecte du XIIIe siècle*, Paris: Imprimerie impériale, 1858. Courtesy of the Division of Rare and Manuscript Collections, Cornell University Library.

Figures 5.1, 5.3, 5.5-5.8, 5.10, 6.14, 7.3, 8.4b, 10.1a, 11.4, 17.1, 18.16, B.3: photos by David W. Henderson of crocheted surfaces designed and executed by Daina Taimiṇa.

Figure 5.4, B.4: photo by David W. Henderson of paper model designed and produced by Keith Henderson (Arnold, Missouri). Used by permission.

Figure 21.1: from Franz van Schooten, *Exercitationum Mathematicicorum, Liber IV.* Leiden (1657), p357. Courtesy of the Division of Rare and Manuscript Collections, Cornell University Library.

Figures 21.2, 21.3, 21.7a, 21.14, 21.15: photos by Francis C. Moon, Professor of Mechanical Engineering, Cornell University: photographs of models in Cornell University's Reuleaux Kinematic Model Collection. Used by permission.

Figure 21.14: drawing added by authors.

All other figures are line drawings, drawn by the authors.

CONTENTS

PREFACE

In mathematics, as in any scientific research, we find two tendencies present. On the one hand, the tendency toward *abstraction* seeks to crystallize the *logical* relations inherent in the maze of material that is being studied, and to correlate the material in a systematic and orderly manner. On the other hand, the tendency toward *intuitive understanding* fosters a more immediate grasp of the objects one studies, a live *rapport* with them, so to speak, which stresses the concrete meaning of their relations.

As to geometry, in particular, the abstract tendency has here led to the magnificent systematic theories of Algebraic Geometry, of Riemannian Geometry, and of Topology; these theories make extensive use of abstract reasoning and symbolic calculation in the sense of algebra. Notwithstanding this, it is still as true today as it ever was that intuitive understanding plays a major role in geometry. And such concrete intuition is of great value not only for the research worker, but also for anyone who wishes to study and appreciate the results of research in geometry.

— David Hilbert [**EG:** Hilbert, p. iii]

These words were written in 1934 by the "father of Formalism" David Hilbert (1862–1943) in the Preface to *Geometry and the Imagination* by Hilbert and S. Cohn-Vossen. Hilbert has emphasized the point we wish to make in this book:

Meaning is important in mathematics and geometry is an important source of that meaning.

We believe that mathematics is a natural and deep part of human experience and that experiences of meaning in mathematics are accessible to everyone. Much of mathematics is not accessible through formal approaches except to those with specialized learning. However, through the use of nonformal experience and geometric imagery, many levels of meaning in mathematics can be opened up in a way that most humans can experience and find intellectually challenging and stimulating.

Formalism contains the power of the meaning but not the meaning. It is necessary to bring the power back to the meaning.

A formal proof as we normally conceive of it is not the goal of mathematics — it is a tool — a means to an end. The goal is understanding. Without understanding we will never be satisfied — with understanding we want to expand that understanding and to communicate it to others. This book is based on a view of proof as a *convincing communication that answers — Why?*

Many formal aspects of mathematics have now been mechanized and this mechanization is widely available on personal computers or even handheld calculators, but the experience of meaning in mathematics is still a human enterprise that is necessary for creative work.

In this book we invite the reader to explore the basic ideas of geometry from a more mature standpoint. We will suggest some of the deeper meanings, larger contexts, and interrelations of the ideas. We are interested in conveying a different approach to mathematics, stimulating the reader to take a broader and deeper view of mathematics and to experience for herself/himself a sense of mathematizing. Through an active participation with these ideas, including exploring and writing about them, people can gain a broader context and experience. This active participation is vital for anyone who wishes to understand mathematics at a deeper level, or anyone wishing to understand something in their experience through the vehicle of mathematics.

This is particularly true for teachers or prospective teachers who are approaching related topics in the school curriculum. All too often we convey to students that mathematics is a closed system, with a single answer or approach to every problem, and often without a larger context. We believe that even where there are strict curricular constraints, there is room to change the meaning and the experience of mathematics in the classroom.

This book is based on a junior/senior-level course that David started teaching in 1974 at Cornell for mathematics majors, high school teachers, future high school teachers, and others. Most of the chapters start intuitively so that they are accessible to a general reader with no particular mathematics background except imagination and a willingness to struggle with ideas. However, the discussions in the book were written for mathematics majors and mathematics teachers and thus assume of the reader a corresponding level of interest and mathematical sophistication.

The course emphasizes learning geometry using reason, intuitive understanding, and insightful personal experiences of meanings in geometry. To accomplish this the students are given a series of inviting and challenging problems and are encouraged to write and speak their reasonings and understandings.

Most of the problems are placed in an appropriate history perspective and approached both in the context of the plane and in the context of a sphere or hyperbolic plane (and sometimes a geometric manifold). We find that by exploring the geometry of a sphere and a hyperbolic plane, our students gain a deeper understanding of the geometry of the (Euclidean) plane.

We introduce the modern notion of "parallel transport along a geodesic," which is a notion of parallelism that makes sense on the plane but also on a sphere or hyperbolic plane (in fact, on any surface). While exploring parallel transport on a sphere, students are able to appreciate more fully that the similarities and differences between the Euclidean geometry of the plane and the non-Euclidean geometries of a sphere or hyperbolic plane are not adequately described by the usual Parallel Postulate. We find that the early interplay between the plane and spheres and hyperbolic planes enriches all the later topics whether on the plane or on spheres and hyperbolic planes. All of these benefits will also exist by only studying the plane and spheres for those instructors that choose to do so.

CHANGES IN THIS EDITION

This book is an expansion and revision of the book *Experiencing Geometry on Plane and Sphere* (1996) and the book *Experiencing Geometry in Euclidean, Spherical, and Hyperbolic Spaces* (2001). There are several important changes: First, there are now co-authors — Daina was a "contributor" to the second edition. She brings considerable experience with and knowledge of the history of mathematics. We start in Chapter 0 with an introduction to four strands in the history of geometry and use the framework of these strands to infuse history into (almost) every chapter in the book in ways to enhance the students understanding and to clear up many misconceptions. There are two new chapters — the old Chapter 14 (Circles in the Plane) has been split into two new chapters: Chapter 15 (on circles with added results on spheres and hyperbolic plane and about trisecting angles and other constructions) and Chapter 16 (on inversions with added material on applications). There is also a new Chapter 21, on the geometry of mechanisms that includes historical machines and results in modern mathematics.

We have included discussions of four new geometric results announced in 2003–2004: In Chapter 12 we describe the discovery and solution of Archimedes' *Stomacion Problem*. Problem **15.2** is based on the 2003 generalization of the notion of power of a point to spheres. In Chapter 16 we talk about applications of a problem of Apollonius to modern pharmacology. In Chapter 18 we discuss the newly announced solution of the Poincaré Conjecture. In Chapter 22 we bring in a new result about unfolding linkages. In Chapter 24 we discuss the latest updates on the shape of space, including the possibility that the shape of the universe is based on a dodecahedron. In addition, we have rearranged and clarified other chapters from the earlier editions.

USEFUL SUPPLEMENTS

We will maintain an Experiencing Geometry Web page:

www.math.cornell.edu/~henderson/ExpGeom/

containing any errata that readers find, updates on new results and discoveries, useful Web references, and other supplementary material.

For exploring properties on a sphere it is important that you have a model of a sphere that you can use. Some people find it helpful to purchase Lénárt Sphere® sets — a transparent sphere, a spherical compass, and a spherical "straightedge" that doubles as a protractor. They work well for small group explorations in the classroom and are available from Key Curriculum Press. However, considerably less expensive alternatives are available: A beach ball or basketball will work for classroom demonstrations, particularly if used with rubber bands large enough to form great circles on the ball. Students often find it convenient to use worn tennis balls ("worn" because the fuzz can get in the way) because they can be written on and are the right size for ordinary rubber bands to represent great circles. Also, many craft stores carry inexpensive plastic spheres that can be used successfully.

If you will be studying the hyperbolic geometry parts of this book, we strongly urge that you have a hyperbolic surface such as those described in Chapter 5. Unfortunately, such hyperbolic surfaces are not readily available commercially. However, directions for making such surfaces (out of paper or by crocheting) are contained in Appendix B, and we will post patterns for making paper models and sources for crocheted hyperbolic surfaces on the Experiencing Geometry Web page as they become available.

The use of commercial dynamic geometry software such as *Geometers Sketchpad*®, *Cabri*®, or *Cinderella*® will enhance any geometry course. These software packages were originally written for exploring Euclidean plane geometry, but recent versions allow one to explore also spherical and hyperbolic geometries. There are also available free on the Web many dynamic geometry programs with more limited functionality. We will maintain on the Experiencing Geometry Web page links to information about these and other software packages and to Web pages that give examples showing their use for self-learning or in a classroom.

A faculty member may obtain from the publisher the *Instructor's Manual* (containing possible solutions to each problem and discussions on how to use this book in a course) by sending a request via e-mail to George_Lobell@prenhall.com or calling 1-201-236-7407.

OUR BACKGROUND IN GEOMETRY

Since an early age David loved to explore geometry in various forms in art classes, woodcarving, carpentry, exploring nature, or by becoming involved in photography. But he did not realize that the geometry, which he experienced, was mathematics or even that it was called "geometry". He was not calling it geometry — he was calling it drawing or design or not calling it anything and just doing it. He did not like mathematics in school because it seemed very dead to him — just memorizing techniques for computing things, and he was not very good at memorizing. He especially did not like his high school geometry course with its formal two-column proofs. This continued on into the university, where David was a joint physics and philosophy major and took only those mathematics courses that were required for physics majors. He became absorbed in geometry-based aspects of physics: mechanics, optics, electricity and magnetism, and relativity. On the other hand, his first mathematics research paper (on the geometry of Venn diagrams for more than four classes) evolved from a course on the philosophy of logic. There were no university geometry courses except for one on "Analytic Geometry and Linear Algebra," which only lightly touched on anything geometric. In his last year as an undergraduate he switched into mathematics because he finally saw that the geometry he loved really was a part of mathematics. In graduate school he studied geometric topology under his mentor, R. H. Bing, who taught without lectures or textbooks in a style that is often known as the *Moore Method*, named after Bing's graduate mentor R. L. Moore at the University of Texas. David joined the faculty at Cornell in 1966 teaching topology and did not start teaching geometry until 1974. David's teaching of the geometry course and the initial writing of this book evolved from this background.

Daina's teaching and interests in geometry and history of mathematics grew out of her experiences in Latvia, which has had a strong tradition of excellence in mathematics at all educational levels. For three years in high school Daina's mathematics teacher was a former university mathematics professor with a passion for geometry — his method of teaching was to give the students large numbers of interesting problems. This teacher always paid a lot of attention to how the students drew geometric diagrams; he encouraged Euclidean constructions with compass and straight edge but also free-hand drawing of geometric figures, insisting on accurate shapes and proportions. At university, Daina took many traditional geometry courses; she always enjoyed and excelled at the drawing aspects of geometry but did not think it had anything to do with art or aesthetic sensibilities. During her graduate years at the university she taught in middle and high school and found that she could raise the students' interest in mathematics by telling them stories from the history of mathematics. She wanted to learn more history, but there was no history of mathematics course at the university. She complained about this to the chair of the department in 1979 and he immediately assigned her to develop and teach one. In 1990 her *History of Mathematics*: *Textbook for University Students* (in Latvian) was published. Daina has taught geometry courses since 1977. Her Ph.D. thesis was in theoretical computer science under the direction of Rusins Freivalds, who still is her model of a true mathematician.

In 1995, David and Daina met in Sicily at an invitation-only ICMI (International Commission on Mathematics Instruction) Study conference, "Teaching Geometry in the 21st Century."

ACKNOWLEDGMENTS FOR THE FIRST EDITION

I acknowledge my debt to all the students and teachers who have attended my geometry courses. Most of these people have been students at Cornell or teachers in the surrounding area of upstate New York, but they also include students at Birzeit University in Palestine and teachers in the new South Africa. Without them this book would have been an impossibility.

Starting in 1986, Avery Solomon and I organized and taught a program of inservice courses for high school teachers under the financial support of Title IIA Grants administered by the New York State Department of Education. This is now called the Cornell/Schools Mathematics Resource Program (CSMRP). As a part of CSMRP, we started recording classes and writing notes on the material. Some of the material in this book had its origins in those notes, but the notes never threatened to become a textbook. I thank Avery for his modeling of enthusiastic teaching, his sharp insights, and his insistence on preserving the teaching materials. In addition to Avery, my friends Marwan Awartani, a professor at Birzeit University, and John Volmink, the director of the Centre for the Advancement of Science and Mathematics Education in Durban, South Africa, have for a long period of time consistently encouraged me to write this book.

A few years ago my colleague Maria Terrell suggested that five of us at Cornell who have been teaching nontraditional geometry courses (Avery Solomon, Bob Connelly, Tom Rishel, Maria, and I) submit a proposal to the National Science Foundation for a grant to write up materials on our courses. The fact that we were awarded the grant (in 1992) is largely due to Maria's persistence, clear thinking, and encouragement. It is this grant that gave me the necessary support to start the writing of this book. I thank the NSF's Program on Course and Curriculum Development for its support.

The major portions of this book were written during the 1992–1993 academic year, in which I taught the course both semesters. Eduarda Moura was my teaching assistant for these courses. She was supported by the NSF grant to assist me by describing the classroom discussion and the student homework on which the content of this book is based. Much of this book (and especially the instructor's manual) is derived from her efforts. In addition to Eduarda, Kelly Gaddis, Beth Porter, Hal Schnee, and Justin Collins were also supported by the grant and made significant contributions to the writing of this book. I thank them all for their excellent contributions, their support of my work, and their friendship. The final writing and the decisions as to what to include and what not to include have all been mine, but they have been based on the foundation started with Avery and the CSMRP materials and continued with Eduarda, Justin, Kelly, Beth, and Hal during 1992–1993.

Since the spring of 1992, the early drafts of the book have been used by me and others at Cornell and 13 other institutions. Various other individuals have worked through the book outside a classroom setting. From these students, instructors, and others I have received encouragement and much valuable feedback that has resulted in what I consider to be a better book. In particular, I want to thank the following persons for giving me feedback and ideas I have used in this final version: David Bray, Douglas Cashing, Helen Doerr, Jay Graening, Christine Kinsey, István Lénárt, Julie Lubell, Richard Pryor, Amanda Cramer and her students, Erica Flapan and her students, Linda Hill and her students, Tim Kurtz and his students, Judy Roitman and her students, Bob Strichartz and his students, and Walter Whitely and his students. Susan Alida spent many hours proofreading and refining the text, and was my consultant on matters of aesthetics.

David W. Henderson, 1995

ACKNOWLEDGMENTS FOR THE SECOND EDITION

I wish to thank the instructors and students (all over the world) who have used the first edition: *Experiencing Geometry on Plane and Sphere*. Their responses were the first encouragement to expand and revise the book.

I wish to thank Jeffrey Weeks for introducing me to the current issues and upcoming experiments about the shape of our physical universe. He is the first to have informed me about the observations that are due to be performed in 2000–2001 that may allow a group of mathematicians and physicists (including Weeks) to determine the global geometry of our physical universe. It is my hope that this book provides the necessary background to understand these observations and determinations.

Without Daina Taimiṇa's crocheted hyperbolic planes I would not have had the intuitive experiences that encouraged me to write this expansion and revision. The ideas for the expansion and revisions were discussed between us and much of the rewriting and expansion was completed with her able assistance.

In addition, the following persons gave me special comments that were incorporated into the expansion and revision: Sarah-Marie Belcastro (U. of Northern Iowa), David Bellamy (U. of Delaware), Gian Mario Besana (Eastern Michigan U.), Alexander Bogomolny (CTK Software), Katherine Borgen (U. of British Columbia), Sean Bradley (Clarke C.), David Dennis (U. of Texas at El Paso), Kelly Gaddis (Lewis and Clarke U.), Paul J. Gies (U. of Maine, Farmington), Chaim Goodman-Strauss (U. of Arkansas), Alice Guckin (C. of Saint Scholastica), Cathy Hayes, Keith Henderson (Thomas Jefferson School, St. Louis), George Litman (National-Louis U.), Jane-Jane Lo (Ithaca C. and Cornell U.), Alan Macdonald (Luther C.), John McCleary (Vassar C.), Nathaniel Miller (Cornell U.), David Mond (U. of Warwick), Colm Mulcahy (Spelman C.), Jodie Novak (U. of Northern Colorado), David A. Olson (MTU), Mary Platt (Salem State C.), John Poland (Carleton C.), Nancy Rodgers (Hanover C.), Thomas Sibley (St. John's U.), Judith Roitman (U. of Kansas), Frances Rosamond (National U.), Avery Solomon (Cornell U.), Daniel Steinberg (Case Western Reserve U.), Robert Stolz (U. of the Virgin Islands), John Sullivan (U. of Illinois, U-C), Margaret Symington (U. of Texas), George Tintera (Texas A&M U., Corpus Christi), Susan Tolman (U. of Illinois, U-C), Andy Vidan (Cornell U.), Tad Watanabe (Towson State U.), Walter Whiteley (York U.), Jeffrey Weeks (Canton, NY), Steve Weissburg (Ithaca High School, Ithaca, NY), Cindy Wyels (California Lutheran U.), and Michelle Zandieh (Arizona State U.). I may have inadvertently left out a few names; if so, I apologize.

David W. Henderson, 2000

ACKNOWLEDGMENTS FOR THIS EDITION

We wish to thank all the people who have participated in our workshops on teaching geometry in the United States, Canada, Latvia, Estonia, and Norway. These workshops have been a great opportunity to interact around the ideas of geometry. In addition, teaching from the Second Edition to students at Cornell and the University of Latvia has given us varied and valuable feedback about what works and what doesn't work.

The following persons gave us special comments that were incorporated into this expansion and revision: Robert Almgren (U. of Toronto), Gian Mario Besana (Indiana U. Northwest), Eric Bray (student, Cornell U.), Diane Bouvier (student, California State U., Long Beach), Tom Braden (U. of Massachusetts), Ken Burns (student, Cornell Unversity), Yuanan Diao (U. of North Carolina at Charlotte), William Dickinson (Grand Valley State University), Barbara Edwards (Oregon State U.), Nir Yehoshua Etzion (student, Cornell U.), Robert Foote (Wabash C.), Kelly Gaddis (Lewis and Clark U.), Catherine Gorini (Maharishi U.), Sarah J. Greenwald (Appalachian State University), Alice Mae Guckin (C. of St. Scholastica), Charlie Jacobson (Elmira C.), Jeff Johannes (SUNY Geneseo), Norman L. Johnson (U. of Iowa), Mushfeq Khan (student, Cornell U.), Ching-chia Ko (student, Oregon State U.), Brian Kowolowski (student, Cornell U.), Wicharn Lewkeeratiyutkul (Chulalongkorn U., Bangkok), Don Lichtenberg (Indiana U.), Anthony M. Lloyd (CHESS, Cornell U.), Jim Loats (Metropolitan State C. of Denver), Robert Lubarsky (Florida Atlantic U.), Philip Mallinson (Phillips Exeter Academy), Douglas Mitarotonda (student, Cornell U.), Margaret Morrow (SUNY Plattsburgh), Karen Mortensen (U. of Illinois, U-C), Michael Mozina, Colm Mulcahy (Spelman C.), G. L. Narasimham (Indian Space Reseach Organization), Derek Rhodes, Laura Ruganis (student, Cornell U.), Konstantin Rybnikov (Cornell U.), Mohammad Salmassi (Framingham C.), Angelo Segalla (California State U., Long Beach), Thomas Sibley (Saint John's U.), Peter Stiller (Texas A&M U.), Hortensia Soto Johnson (Colorado State U., Pueblo), Phil Sweet (student, SUNY Plattsburgh), Steve Weissburg (Ithaca High School, Ithaca, NY), John Sullivan (U. of Illinois, U-C), Jeffrey Weeks (Canton, NY), Walter Whiteley (York U.). Thanks to Frank Moon (Mechanical Engineering, Cornell U.) for giving us permission to use his photographs.

We produced the entire manuscript (typing, formatting, drawings, and final layout) using the integrated word processing software Word-Pro. Finally, we wish to thank George Lobell, Senior Editor at Prentice Hall, for his encouragement and support and for the vision and enthusiasm with which he shepherded all three editions through the publication process.

David W. Henderson
henderson@math.cornell.edu
Daina Taimiņa
dtaimina@math.cornell.edu

How to Use This Book

Do not just pay attention to the words;
Instead pay attention to meanings behind the words.
But, do not just pay attention to meanings behind the words;
Instead pay attention to your deep experience of those meanings.
— Tenzin Gyatso, The Fourteenth Dalai Lama
From an unpublished lecture in London, April 1984.

This quote expresses the philosophy on which this book is based. Most of the chapters start intuitively so that they are accessible to a general reader with no particular mathematics background except imagination and a willingness to struggle with one's own experience of the meanings. However, the discussions in the book were written for mathematics majors and mathematics teachers and thus assume of the reader a corresponding level of interest and mathematical sophistication.

This book will present you with a series of problems. You should explore each question and write out your thinking in a way that can be shared with others. By doing this you will be able to develop ideas actively prior to passively reading or listening to comments of others. When working on the problems, you should be open-minded and flexible and let your thinking wander. Some problems will have short, fairly definitive answers, and others will lead into deep areas of meaning that can be probed almost indefinitely. You should not accept anything just because you remember it from school or because some authority says it's good. Insist on understanding (or seeing) why it is true or what it means for you. Pay attention to *your* deep experience of these meanings.

You should think about each problem and express your thinking even when you know you cannot complete it. This is important because

- ◆ It helps build self-confidence.
- ◆ You will see what your real difficulties are.
- ◆ When you see a solution or proof later, you will more likely see it as answering a question that you have.

An important thing to keep in mind is that there is no one correct solution. There are many different ways of solving the problems — as many as there are ways of understanding the problems. *Insist on understanding* (or seeing) why it is true or what it means to you. Everyone understands things in a different way, and one person's "obvious" solution may not work for you. However, it is helpful to talk with others — listen to their ideas and confusions and then share your ideas and confusions with them. In the experience of those using this book and the earlier edition, it is very important to be able to talk with others in small groups whether inside or outside of class. In fact, small groups have successfully gone through this book as self-study without a teacher.

Also, some of the problems are difficult to visualize in your head. Make models, draw pictures, use rubber bands on a ball, use scissors and paper — play!

For exploring properties on a sphere it is important that you have a model of a sphere that you can use. You can draw on worn tennis balls ("worn" because the fuzz can get in the way) and they are the right size for ordinary rubber bands to represent great circles. You may find useful clear plastic spheres from craft stores. Most any ball you have around will work — you can even use an orange and then eat it when you get hungry!

For exploring the geometric properties of a hyperbolic plane it is very important to have a hyperbolic surface in your hands. Instructions on how you can make (either out of paper or by crocheting) hyperbolic surfaces are contained in the beginning of Chapter 5. It will be very helpful to your understanding of the hyperbolic plane for you to actually make one of these hyperbolic surfaces yourself.

HOW WE USE THIS BOOK IN A COURSE

In our course *the distinction between learning activities and assessment activities is blurred*. We present a sequence of problems (together with motivation, discussion of contexts, and connections of the problems with other areas of mathematics and life). We tell the students the following:

Write out your thinking to each problem. We will return your papers with comments about your solutions. Respond to our comments — use them as invitations to explore, to clarify your understanding of the problem, or to clarify our understanding of your solution. Keep responding until you understand. Turn in whatever your thinking is on a question even if only to say "I don't understand such and such" or "I'm stuck here"; be as specific as possible. Feel free to ask questions. This will allow the sharing of ideas, and you will benefit more from class sessions.

The students then work on the problems either individually or in small groups and report their thinking back to us and the class. This cycle of ***writing, comments, discussion*** continues on each problem until both the students and we are satisfied, unless external constraints of time and resources intervene.

Some problems are investigated by our students in small cooperative learning groups (with no written work) and the groups report back to the class, or not, depending on how it goes.

What we have discovered is that in this process not only have the students learned from the course, but also we have learned much about geometry from them. At first we were surprised; how could the teacher learn mathematics from the students? But this learning has continued for 25 years and we now expect its occurrence. For more discussion of this, see David's article "I Learn Mathematics from My Students — Multiculturalism in Action," *For the Learning of Mathematics*, 16(1996), 34–40.

For a final project we usually ask each student to pick a chapter that we have not covered in class and to investigate it on his or her own (with our assistance as necessary in office hours). We have found from the experiences of our students that all the chapters work for such projects.

We do not teach in the same way. The most significant difference is that when students in Daina's classes get the main point of a problem, then they move on, while in David's classes the students spend more time on each problem exploring it for as much material as possible. Thus Daina's students cover more problems than do David's students.

BUT DO IT YOUR OWN WAY

From feedback we have received from instructors using the first two editions of this book, we have learned that many of the most successful uses of this book in the classroom are when the instructor does not follow this book! In many of these successful courses, the instructors have used some of the problems from this book and then added their own favorite problems. The chapters in the book have been used in a variety of different orders and at many different speeds. Some instructors have, with careful organization and direction, successfully covered the whole of the first edition of this book in a semester. Other instructors have let their students explore and wander in such a way that they covered less than half of the book. Some instructors use the book in a course with traditional preliminary and final exams. Others have had no exams but instead relied on portfolios, student journals, and/or projects. Some have used small cooperative learning groups regularly throughout the semester and others have never divided the students into small groups.

Also, the book has been used in courses for sophomores who are prospective teachers — either elementary or secondary; in courses for senior mathematics majors; in courses for Master's students in Education. Many of the problems in this book (but not the book itself) have been used successfully with freshman-level courses for students weak in mathematics ("liberal arts mathematics"). Portions of the book have also been used successfully in a freshman writing course.

CHAPTER SEQUENCES

Minimum Core: Cover, at least, Chapters 0–3, 6, 8, and either 7 or 10.

After exploring the Minimum Core, all of the other chapters are essentially independent: Each chapter refers to some results from other chapters but these references are indicated and can be looked up without destroying the experience of the chapter you are exploring.

Starred problems and sections: Certain problems and sections in this book require from the reader a more advanced background — these sections are indicated with an asterisk (*). These problems and sections may be omitted without seriously affecting the later explorations.

Here are some chapter sequences that put emphasis on different aspects of geometry:

To use this book the same as the first edition: Leave out Chapters 5, 15–18, 21, and 24; and ignore the references to hyperbolic geometry and the hyperbolic plane in the remaining chapters.

To emphasize traditional topics in Euclidean plane geometry: Explore Chapters 1–3, 6, 8–13, 15, 16, 19, and 23. This includes some geometry on spheres, which is needed to bring out the Euclidean meanings; however, you can leave out all mentions of hyperbolic geometry.

To emphasize spherical geometry: Explore Chapters 1–3, 6–9, 11, 12, 14, 20, 22, and 23.

To emphasize hyperbolic geometry: Explore Chapters 1–3, and the hyperbolic plane parts of Chapters 5–12, 16, and 17.

To get to Chapter 24 on the shape of space as quickly as possible: Explore Chapters 1–8, and at least Problems **18.1**, **18.6**, **22.1**, and **22.6**.

For chapters (14–20) using similar triangles: Explore Problems **12.1** and **13.3** or assume the Criteria for Similar Triangles (**13.4**).

For three-dimensional investigations in Chapters 23 and 24: At least explore Problem **22.1** and assume the results of Problem **22.6**.

These chapter sequences are summarized in the following table:

Goal	Chapters																
Minimum Core	1–3, 6, 8, and either 7 or 10																
Use book similar to first edition	4		7	9	10	11	12	13	14					19	20	22	23
To emphasize Euclidean geometry				9	10	11	12	13		15	16			19			23
To emphasize spherical geometry			7	9		11	12		14						20	22	23
To emphasize hyperbolic geometry		5	7	9	10	11	12				16	17					
To get to Chapter 24 "Shape of Space" as soon as possible	4	5	7										18			22.1 & 22.6	
To prove and apply similar triangle criteria (**13.4**)				9	10		12	13	14	15	16	17	18	19	20		

Professors experiencing geometry in a workshop

Chapter 0

HISTORICAL STRANDS OF GEOMETRY

All people by nature desire knowledge.
> — Aristotle (384 B.C.–322 B.C.), *Metaphysics*

History is the witness that testifies to the passing of time; it illumines reality, vitalizes memory, provides guidance in daily life and brings us tidings of antiquity.
> — Cicero (106 B.C.–43 B.C.), *Pro Publio Sestio*

Inherited ideas are a curious thing, and interesting to observe and examine.
> — Mark Twain (1835–1910)
> *A Connecticut Yankee in King Arthur's Court*

We think that the main aspects of geometry today emerged from four strands of early human activity that seem to have occurred in most cultures: art/pattern, building structures, motion/machines, and navigation/stargazing. These strands developed more or less independently into varying studies and practices that eventually from the 19th century on were woven into what we now call *geometry*.

ART/PATTERN STRAND

To produce decorations for their weaving, pottery, and other objects, early artists experimented with symmetries and repeating patterns. See Figure 0.1. Later the study of symmetries of patterns led to tilings, group

1

theory, crystallography, and finite geometries. See examples in Chapters 1, 11, 18, and 24.

Figure 0.1   Ancient Greek geometric art with examples of strip patterns
(Chapter 11)

Figure 0.2   *Giving of the Keys to Saint Peter* by Pietro Perugina using perspective
(Chapter 20)

Early artists also explored various methods of representing physical objects, and living things. These explorations in Rennaisance (see Figure 0.2) led to the study of perspective and then later to projective geometry and descriptive geometry. See examples in Problems **17.6** and **20.6**.

In the last two centuries this strand has led to security codes, digital picture compression, computer-aided graphics, the study of computer vision in robotics, and computer-generated movies.

For more details on the Art/Pattern Strand, see [**AD**: Albarn], [**GC**: Bain], [**EM**: Devlin], [**GC**: Eglash], [**AD**: Field], [**GC**: Gerdes], [**AD**: Ghyka], [**AD**: Gombrich], [**SG**: Hargittai], [**AD**: Ivins], [**AD**: Kappraf].

NAVIGATION/STARGAZING STRAND

Figure 0.3 Armilary sphere based on Greeks' view of the universe (Chapter 2)

For astrological, religious, agricultural, and other purposes, ancient humans attempted to understand the movement of heavenly bodies (stars, planets, sun, and moon) in the apparently hemispherical sky. See Figure 0.3. Early humans used the stars and planets as they started navigating over long distances on the earth and on the sea. They used this understanding to solve problems in navigation and in attempts to understand the shape of the earth. Ideas of trigonometry apparently were first developed by Babylonians in their studies of the motions of heavenly bodies. Even Euclid wrote an astronomical work, *Phaenomena*, [**AT**: Berggren], in which he studied properties of curves on a sphere.

Navigation and large-scale surveying developed over the centuries around the world and along with it cartography, trigonometry, and spherical geometry. This strand is represented wherever intrinsic geometry of surfaces is discussed, especially Chapters 4, 7, 8, and 18. Map making is discussed in Chapter 16 and trigonometry in Chapter 20.

Examples most closely associated with this strand in the last two centuries are the study of surfaces and manifolds, which led to many modern spatial theories in physics and cosmology. See Figure 0.4 and Chapters 2, 5, 18, 22, and 24.

Figure 0.4   Picture of the cosmic background radiation taken by NASA's WMAP satellite, released February 11, 2003 (Chapter 24)

For more details on the Navigation/Stargazing Strand, see [**CE**: Bagrow], [**HI**: Burkert], [**UN**: Ferguson], [**UN**: Osserman], [**SP**: Todhunter].

BUILDING STRUCTURES STRAND

As humans built shelters, altars, bridges, and other structures, they discovered ways to make circles of various radii, and various polygonal/polyhedral structures. See Figure 0.5. In the process they devised systems of measurement and tools for measuring. The (2000–600 B.C.) *Sulbasutram* [**AT**: Baudhayana] is written for altar builders and contains at the beginning a geometry handbook with proofs of some theorems and a clear general statement of the "Pythagorean" Theorem, (see Chapter 13). Building upon geometric knowledge from Babylonian, Egyptian, and early Greek builders and scholars, Euclid (325–265 B.C.) wrote his *Elements*, which became the most used mathematics textbook in the world for the next 2300 years and codified what we now call Euclidean geometry.

Figure 0.5 Step pyramid of Memphis, Egypt (about 3000 B.C.)

Using the *Elements* as a basis in the period 300 B.C. to about 1000 A.D., Greek and Islamic mathematicians extended its results, refined its postulates, and developed the study of conic sections and geometric algebra. Within Euclidean geometry, analytic geometry, vector geometry (linear algebra and affine geometry), and algebraic geometry developed later. The *Elements* also started what became known as the axiomatic method in mathematics. For the next 2000 years mathematicians attempted to prove Euclid's Fifth (Parallel) Postulate as a theorem (based on the other postulates); these attempts culminated around 1825 with the discovery of hyperbolic geometry. See Appendix A and Chapter 10 for a discussion of Euclid's postulate and various other parallel postulates. Chapters 3, 6, 9, 12, 15, 16, 19, and 23 contain many of the basic topics of Euclidean geometry.

Further developments with axiomatic methods in geometry led to the axiomatic theories of the real numbers and analysis and to elliptic geometries, axiomatic projective geometry, and other axiomatic geometries. See the three sections at the end of Chapter 10 for more discussion. And, of course, Euclidean geometry has continued to be used in the building of modern structures (for example, Figure 0.6).

Figure 0.6   Apartment building in Baltimore

For more detail on the Building Structures Strand, see [**AT**: Baudhayana], [**AD**: Blackwell], [**HI**: Burkert], [**GC**: Datta], [**ME**: DeCamp], [**TX**: Hartshorne], [**HI**: Heilbron], [**HI**: Seidenberg].

MOTION/MACHINES STRAND

Early human societies used the wheel for transportation, for making pottery, in pulleys, and in pumps.

In ancient Greece, Archimedes (see Figure 0.7), Hero, and other geometers used linkages and gears to solve geometrical problems — such as trisecting an angle, duplicating a cube, and squaring a circle. See Problem **15.4**. These solutions were not accepted in the Building Structures Strand, and this leads to a common misconception that these problems are unsolvable and/or that Greeks did not allow motion in geometry. See [**TX**: Martin] and [**HI**: Katz].

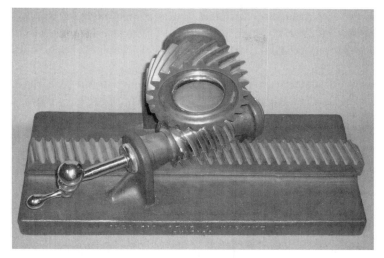

Figure 0.7 Endless screw such as used by Archimedes and da Vinci

We do not know much about the influence of machines on mathematics after ancient Greece until the advent of machines in the Renaissance. In the 17th century, Descartes and other mathematicians used linkages to study curves. This study of curves led to the development of analytic geometry. Mechanical computing machines were also developed in the 17th century.

Figure 0.8 Wankel engine based on a Reuleaux triangle (Chapter 21)

As we will discuss in Chapter 21, there was an interaction between mathematics and mechanics that led to marvelous machine design (Figure 0.8) and continues to the modern mathematics of rigidity and robotics.

For more details on the Motion/Machines Strand, see [**ME**: De-Camp], [**ME**: Dyson], [**ME**: Ferguson 2001], [**ME**: Kirby], [**ME**: Moon], [**ME**: Ramelli], [**ME**: Williams].

Chapter 1

WHAT IS STRAIGHT?

Straight is that of which the middle is in front of both extremities.

— Plato, *Parmenides,* 137 E [**AT**: Plato]

A *straight line* is a line that lies symmetrically with the points on itself.

— Euclid, *Elements*, Definition 4 [Appendix A]

HISTORY: HOW CAN WE DRAW A STRAIGHT LINE?

When using a compass to draw a circle, we are not starting with a model of a circle; instead we are using a fundamental property of circles that the points on a circle are a fixed distance from a center. Or we can say we use Euclid's definition of a circle (see Appendix A, Definition 15). So now what about drawing a straight line: Is there a tool (serving the role of a compass) that will draw a straight line? One could say we can use a straightedge for constructing a straight line. Well, how do you know that your straightedge is straight? How can you check that something is straight? What does "straight" mean? Think about it — this is part of Problem **1.1** below.

You can try to use Euclid's definition above. If you fold a piece of paper the crease will be straight — the edges of the paper needn't even be straight. This utilizes mirror symmetry to produce the straight line. Carpenters also use symmetry to determine straightness — they put two boards face to face, plane the edges until they look straight, and then turn one board over so the planed edges are touching. See Figure 1.1. They then hold the boards up to the light. If the edges are not straight, there will be gaps between the boards where light will shine through. In Problem **1.1** we will explore more deeply symmetries of a straight line.

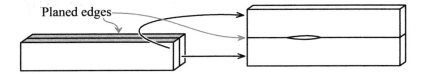

Figure 1.1 Carpenter's method for checking straightness

When grinding an extremely accurate flat mirror, the following technique is sometimes used: Take three approximately flat pieces of glass and put pumice between the first and second pieces and grind them together. Then do the same for the second and the third pieces and then for the third and first pieces. Repeat many times and all three pieces of glass will become very accurately flat. See Figure 1.2. Do you see why this works? What does this have to do with straightness?

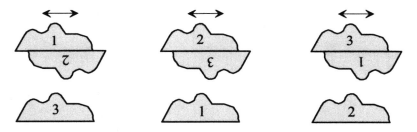

Figure 1.2   Grinding flat mirrors

We can also use the usual high school definition, "A straight line is the shortest distance between two points." This leads to producing a straight line by stretching a string.

This use of symmetry, stretching, and folding can also be extended to other surfaces, as we will see in Chapters 2, 4, and 5. Sometimes we can get confused when reading in the literature that "straight line" is an undefined term or that straight lines on the sphere are "defined to be arcs of great circles." We find that putting "straight" in the context of the four historical strands helps clarify this: "Symmetry" comes mostly from the Art/Pattern Strand, "undefined term" comes from the axiomatics in the Building Structures Strand, and "shortest distance" comes mostly from the Navigation/Stargazing Strand.

But there is still left unanswered the question of whether there is a mechanism analogous to a compass that will draw an accurate straight line. We can find answers to this question in the history of mechanics, which leads us into the Motion/Machines Strand and to another meaning of "straight".

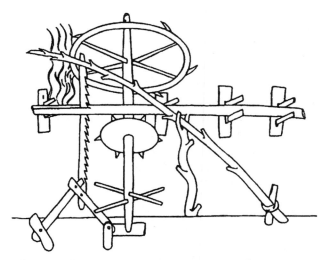

Figure 1.3 Up-and-down sawmill of the 13th century

Turning circular motion into straight line motion has been a practical engineering problem since at least the 13th century. As we can see in some 13th century drawings of a sawmill (see Figure 1.3) linkages (rigid bars constrained to be near a plane and joined at their ends by rivets) were in use in the 13th century and probably were originated much earlier. Georgius Agricola's (1494–1555) geological writings [**ME**: Agricola] reflect firsthand observations not just of rocks and minerals, but of every aspect of mining technology and practice of the time. In the pictures of his work one can see link work that was widely used for converting the continuous rotation of a water wheel into a reciprocating motion suited to piston pumps. In 1588, Agostino Ramelli published his book [**ME**: Ramelli] on machines where linkages were widely used. Both of these books are readily available; see the Bibliography.

In the late 18th century people started turning to steam engines for power. James Watt (1736–1819), a highly gifted designer of machines, worked on improving the efficiency and power of steam engines. In a steam engine, the steam pressure pushes a piston down a straight cylinder. Watt's problem was how to turn this linear motion into the circular motion of a wheel (such as on steam locomotives). It took Watt several years to design the straight-line linkage that would change straight-line motion to circular one. Later, Watt told his son,

> Though I am not over anxious after fame, yet I am more proud of the parallel motion than of any other mechanical invention I have ever made. (quoted in [**ME**: Ferguson 1962], p 197])

"Parallel motion" is a name Watt used for his linkage, which was included in an extensive patent of 1784. Watt's linkage was a good solution to the practical engineering problem. See Figure 1.4, where his linkage is the parallelogram and associated links in the upper left corner.

Figure 1.4 A steam engine with Watt's "parallel motion" linkage

But Watt's solution did not satisfy mathematicians, who knew that his linkage can draw only an approximate straight line. Mathematicians continued to look for a planar straight-line linkage. A linkage that would draw an exact straight-line was not found until 1864–1871, when a French army officer, Charles Nicolas Peaucellier (1832–1913), and a Russian graduate student, Lipmann I. Lipkin (1851–1875), independently developed a linkage that draws an exact straight line. See Figure 1.5. (There is not much known about Lipkin. Some sources mentioned that he was born in Lithuania and was a graduate student of Chebyshev in Saint Petersburg but died before completing his doctoral dissertation.) (For more discussion of this discovery see also Philip Davis's delightful little book *The Thread* [**EM**: Davis], Chapter IV.)

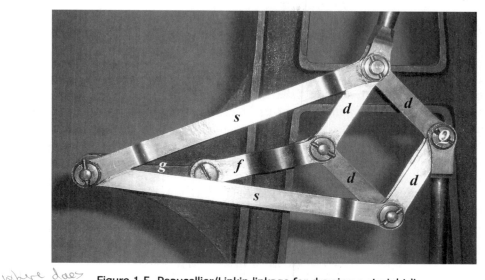

where does this come from

Figure 1.5 Peaucellier/Lipkin linkage for drawing a straight line

The linkage in Figure 1.5 works because, as we will show in Problem **16.3**, the point Q will only move along an arc of a circle of radius $(s^2 - d^2)f/(g^2 - f^2)$. This allows one to draw an arc of a large circle without using its center. When the lengths g and f are equal, then P draws the arc of a circle with infinite radius. (See [**EG**: Hilbert], pp. 272–273, for another discussion of this linkage.) So we find in the Motion/Machines Strand another notion of straight line as a circle of infinite radius (see the text near Figure 11.4 for discussion of circles of infinite radius).

PROBLEM 1.1 WHEN DO YOU CALL A LINE STRAIGHT?

In keeping with the spirit of the approach to geometry discussed in the Preface, we begin with a question that encourages you to explore deeply the concept of straightness. We ask you to build a notion of straightness from your experiences rather than accept a certain number of assumptions about straightness. Though it is difficult to formalize, straightness is a natural human concept.

   a.   *How can you check in a practical way if something is straight? How do you construct something straight — lay out fence posts in a straight line, or draw a straight line? Do this without assuming that you have a ruler, for then we will ask, "How can you check that the ruler is straight?"*

At first, look for examples of straightness in your experiences. Go out and actually try walking along a straight line and then along a curved path; try drawing a straight line and then checking that a line already drawn is straight. As you look for properties of straight lines that distinguish them from non-straight lines, you will probably remember the following statement (which is often taken as a definition in high school geometry): *A straight line is the shortest distance between two points.* But can you ever measure the lengths of all the paths between two points? How do you find the shortest path? If the shortest path between two points is in fact a straight line, then is the converse true? Is a straight line between two points always the shortest path? We will return to these questions in later chapters.

A powerful approach to this problem is to think about lines in terms of symmetry. This will become increasingly important as we go on to other surfaces (spheres, cones, cylinders, and so forth). Two of the symmetries of lines are as follows:

◆ *Reflection symmetry in the line*, also called *bilateral symmetry* — reflecting (or mirroring) an object over the line.

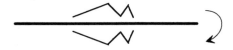

Figure 1.6   Reflection symmetry of a straight line

◆ *Half-turn symmetry* — rotating 180° about any point on the line.

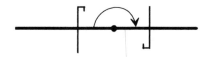

Figure 1.7   Half-turn symmetry of a straight line

Although we are focusing on a symmetry *of the line* in each of these examples, notice that the symmetry is not a property of the line by itself but includes the line and the space around the line. The symmetries preserve the local environment of the line. Notice how in reflection and half-turn symmetry the line and its local environment are both part of the symmetry action and how the relationship between them is integral to the action. In fact, reflection in the line does not move the line at all but exhibits a way in which the spaces on the two sides of the line are the same.

 DEFINITIONS. An *isometry* is a transformation that preserves distances and angle measures. A *symmetry of a figure* is an isometry of a region of space that takes the figure (or the portion of it in the region) onto itself. You will show in Problem **11.3** that every isometry of the plane is either a translation, a rotation, a reflection, or a composition of them.

b. *What symmetries does a straight line have?*

Try to think of other symmetries of a line as well (there are quite a few). Some symmetries hold only for straight lines, while some work for other curves as well. Try to determine which symmetries are specific to straight lines and why. Also think of practical applications of these symmetries for constructing a straight line or for determining if a line is straight.

c. *What is in common among the different notions of straightness? Can you write a definition of "straight line"?*

Look for things that you call "straight." Where do you see straight lines? Why do you say they are straight? Look for both physical lines and non-physical uses of the word "straight". What symmetries does a straight line have? How do they fit with the examples that you have found and those mentioned above? Can we use any of the symmetries of a line to define straightness? The intersection of two (flat) planes is a straight line — why does this work? Does it help us understand "straightness"?

Imagine (or actually try!) walking while pulling a long thread with a small stone attached. When will the stone follow along your path? Why? This property is used to pick up a fallen water skier. The boat travels by the skier along a straight line and thus the tow rope follows the path of the boat. Then the boat turns in an arc in front of the skier. Because the boat is no longer following a straight path, the tow rope moves in toward the fallen skier. What is happening?

Another idea to keep in mind is that straightness must be thought of as a local property. Part of a line can be straight even though the whole line may not be. For example, if we agree that this line is straight,

and then we add a squiggly part on the end, like this,

would we now say that the original part of the line is not straight, even though it hasn't changed, only been added to? Also note that we are not making any distinction here between "line" and "line segment." The more generic term "line" generally works well to refer to any and all lines and line segments, both straight and non-straight.

You are likely to bring up many ideas of straightness. It is necessary then to think about what is common among all of these straight phenomena.

Think about and formulate some answers for these questions before you read any further. Do not take anything for granted unless you see why it is true. No answers are predetermined. You may come up with something that we have never imagined. Consequently, it is important that you persist in following your own ideas. Reread the section "How to Use This Book" starting on page xxv.

You should not read further until you have expressed your thinking and ideas through writing or talking to someone else.

THE SYMMETRIES OF A LINE

Reflection-in-the-line symmetry: It is most useful to think of reflection as a "mirror" action with the line as an axis rather than as a "flip-over" action that involves an action in 3-space. In this way we can extend the notion of reflection symmetry to a sphere (the flip-over action is not possible on a sphere). Notice that this symmetry cannot be used as a definition for straightness because we use straightness to define reflection symmetry. This same comment applies to most of the other symmetries discussed below.

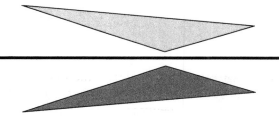

Figure 1.8   Reflection-in-the-line symmetry

In Figures 1.8–1.14, the light gray triangle is the image of the dark gray triangle under the action of the symmetry on the space around the line.

- *Practical application*:   We can produce a straight line by folding a piece of paper because this action forces symmetry along the crease. Above we showed a carpenter's example.

Reflection-perpendicular-to-the-line symmetry: A reflection through *any* axis perpendicular to the line will take the line onto itself. Note that circles also have this symmetry about any diameter. See Figure 1.9.

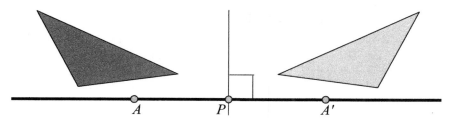

Figure 1.9   Reflection-perpendicular-to-the-line symmetry

- *Practical applications*: You call tell if a straight segment is perpendicular to a mirror by seeing if it looks straight with its reflection. Also, a straight line can be folded onto itself.

✳ **Half-turn symmetry:** A rotation through half of a full revolution about any point *P* on the line takes the part of the line before *P* onto the part of the line after *P* and vice versa. Note that some non-straight lines, such as the letter Z, also have half-turn symmetry — but not about *every* point. See Figure 1.10.

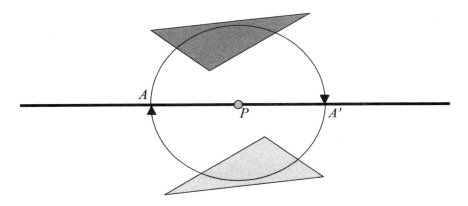

Figure 1.10   Half-turn symmetry

- *Practical applications*: Half-turn symmetry exists for the slot on a screw and the tip of the screwdriver (unless you are using Phillips-head screws and screwdrivers, which also have quarter-turn symmetry) and thus we can more easily put the tip of the screwdriver into the slot. Also, this symmetry is involved when a door (in a straight wall) opens up flat against the wall.

✳ **Rigid-motion-along-itself symmetry:** For straight lines in the plane, we call this *translation symmetry*. Any portion of a straight line may be moved along the line without leaving the line. This property of being able to move rigidly along itself is not unique to straight lines; circles (rotation symmetry) and circular helixes (screw symmetry) have this property as well. See Figure 1.11.

- *Practical applications*: Slide joints such as in trombones, drawers, nuts and bolts, and so forth, all utilize this symmetry.

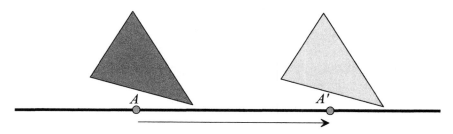

Figure 1.11   Rigid-motion-along-itself symmetry

✳   **3-dimensional-rotation symmetry:** In a 3-dimensional space, rotate the line around itself through *any* angle using itself as an axis.

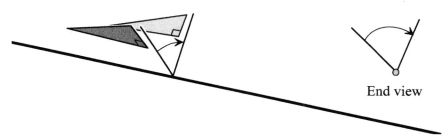

End view

Figure 1.12   3-dimensional-rotation symmetry

- *Practical applications*: This symmetry can be used to check the straightness of any long thin object such as a stick by twirling the stick with itself as the axis. If the stick does not appear to wobble, then it is straight. This is used for pool cues, axles, hinge pins, and so forth.

✳   **Central symmetry or point symmetry:** Central symmetry through the point P sends any point A to the point on the line determined by A and P at the same distance from P but on the opposite side from P. See Figure 1.13. In two dimensions central symmetry does not differ from half-turn symmetry in its end result, but they do differ in the ways we imagine them and construct them.

♦ In 3-space, central symmetry produces a result different than any single rotation or reflection (though we can check that it does give the same result as the composition of three reflections through mutually perpendicular planes). To experience central symmetry in 3-space, hold your hands in front of you with the palms facing each other and your left thumb up and your right thumb down. Your two hands now have approximate central symmetry about a point midway between the center of the palms; and this symmetry cannot be produced by any reflection or rotation.

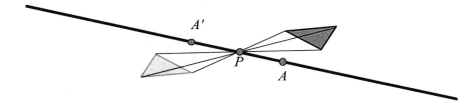

Figure 1.13   Central symmetry

※ **Similarity or self-similarity "quasi-symmetry":** Any segment of a straight line (and its environs) is similar to (that is, can be magnified or shrunk to become the same as) any other segment. See Figure 1.14. This is not a symmetry because it does not preserve distances but it could be called a "quasi-symmetry" because it *does* preserve the measure of angles.

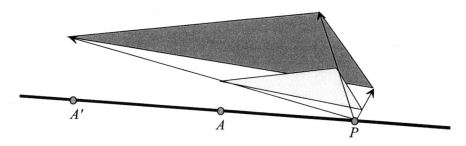

Figure 1.14   Similarity "quasi-symmetry"

♦ Logarithmic spirals such as the chambered nautilus have self-similarity as do many fractals. (See example in Figure 1.15.)

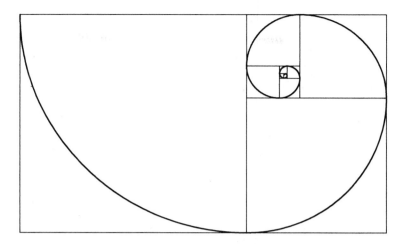

Figure 1.15   Logarithmic spiral

Clearly, other objects besides lines have some of the symmetries mentioned here. It is important for you to construct your own such examples and attempt to find an object that has all of the symmetries but is not a line. This will help you to experience that straightness and the seven symmetries discussed here are intimately related. You should come to the conclusion that while other curves and figures have some of these symmetries, only straight lines have all of them.

LOCAL (AND INFINITESIMAL) STRAIGHTNESS

Previously, you saw how a straight line has reflection-in-the-line symmetry and half-turn symmetry: One side of the line is the same as the other. But, as pointed out above, straightness is a local property in that whether a segment of a line is straight depends only on what is near the segment and does not depend on anything happening away from the line. Thus each of the symmetries must be able to be thought of (and experienced) as applying only locally. This will become particularly important later when we investigate straightness on the cone and cylinder. (See the discussions in Chapter 4.) For now, it can be experienced in the following way:

When a piece of paper is folded not in the center, the crease is still straight even though the two sides of the crease on the paper are not the same. (See Figure 1.16.)

So what is the role of the sides when we are checking for straightness using reflection symmetry? Think about what is important near the crease in order to have reflection symmetry.

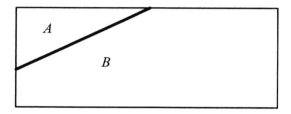

Figure 1.16   Reflection symmetry is local

When we talk about straightness as a local property, you may bring out some notions of scale. For example, if you see only a small portion of a very large circle, it will be indistinguishable from a straight line. This can be experienced easily on many of the modern graphing programs for computers. Also, a microscope with a zoom lens will provide an experience of zooming. If a curve is smooth (or differentiable), then if we "zoom in" on any point of the curve, eventually the curve will be indistinguishable from a straight line segment. See Figure 1.17.

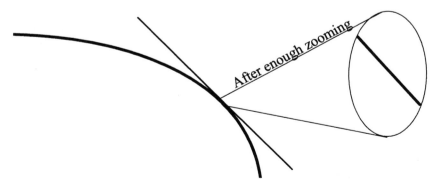

Figure 1.17   Infinitesimally straight

We sometimes use the terminology, *infinitesimally straight*, in place of the more standard terminology, *differentiable*. We say that a curve is *infinitesimally straight* at a point *p* if there is a straight line *l* such that if we zoom in enough on p, the line and the curve become indistinguishable.[1] When the curve is parametrized by arc length this is equivalent to the curve having a well-defined velocity vector at each point.

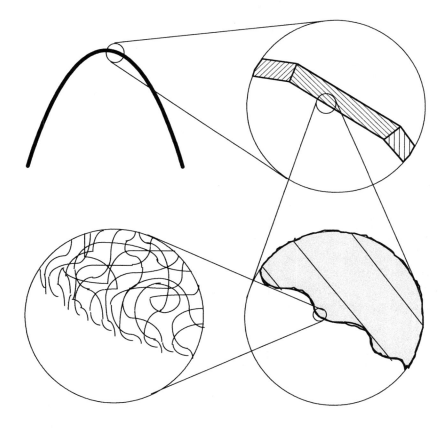

Figure 1.18   Straightness and smoothness depend on the scale

In contrast, we can say that a curve is *locally straight at a point* if that point has a neighborhood that is straight. In the physical world the usual use of both *smooth* and *locally straight* is dependent on the scale at which they are viewed. For example, we may look at an arch made out of wood — at a distance it appears as a smooth curve; then as we move in closer we see that the curve is made by many short straight pieces of finished (planed) boards, but when we are close enough to touch it, we see that its surface is made up of smooth waves or ripples, and under a microscope we see the non-smoothness of numerous twisting fibers. See Figure 1.18.

---

[1] This is equivalent to the usual definitions of being differentiable at p. For example, if $t(x) = f(p) + f'(p)(x - p)$ is the equation of the line tangent to the curve $(x, f(x))$ at the point $(p, f(p))$, then, given $\varepsilon > 0$ (the distance of indistinguishability), there is a $\delta > 0$ (the radius of the zoom window) such that, for $|x - p| < \delta$ (for x within the zoom window), $|f(x) - t(x)| < \varepsilon$ [$f(x)$ is indistinguishable from $t(x)$]. This last inequality may look more familiar in the form

$$f(x) - t(x) = f(x) - f(p) - f'(p)(x - p) = \{ [f(x) - f(p)]/(x - p) - f'(p) \}(x - p) < \varepsilon.$$

In general, the value of δ might depend on p as well as on ε. Often the term *smooth* is used to mean continuously differentiable, which the interested reader can check is equivalent (on closed finite intervals) to, for each $\varepsilon > 0$, there being one $\delta > 0$ that works for all p.

Chapter 2

STRAIGHTNESS ON SPHERES

> ... [I]t will readily be seen how much space lies between the two
> places themselves on the circumference of the large circle which is
> drawn through them around the earth. ... [W]e grant that it has been
> demonstrated by mathematics that the surface of the land and water is
> in its entirety a sphere, ... and that any plane which passes through
> the center makes at its surface, that is, at the surface of the earth and
> of the sky, great circles, and that the angles of the planes, which
> angles are at the center, cut the circumferences of the circles which
> they intercept proportionately, ...
> — Ptolemy, *Geographia* (ca. 150 A.D.) Book One, Chapter II

This chapter asks you to investigate the notion of straightness on a
sphere, drawing on the understandings about straightness you developed
in Problem **1.1**.

EARLY HISTORY OF SPHERICAL GEOMETRY

Observations of heavenly bodies were carried out in ancient Egypt and
Babylon, mainly for astrological purposes and for making a calendar,
which was important for organizing society. Claudius Ptolemy (c.
100–178), in his *Almagest*, cites Babylonian observations of eclipses and
stars dating back to the 8th century B.C. The Babylonians originated the
notion of dividing a circle into 360 degrees — speculations as to why
360 include that it was close to the number of days in a year, it was
convenient to use in their hexadecimal system of counting, and 360 is
the number of ways that seven points can be placed on a circle without
regard to orientation (for the ancients there were seven "wandering
bodies" — sun, moon, Mercury, Venus, Mars, Saturn, and Jupiter). But,
more important, the Babylonians developed a coordinate system

25

(essentially the same as what we now call "spherical coordinates") for the celestial sphere (the apparent sphere on which the stars, sun, moon, and planets appear to move) with its pole at the north star. Thus it is a misconception to think that the use of coordinates originated with Descartes in the 17th century.

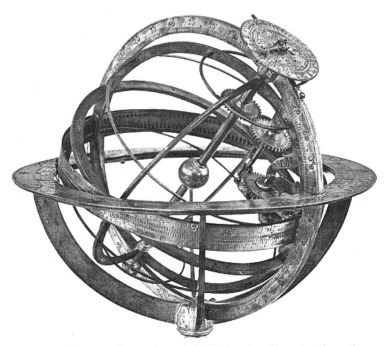

Figure 2.1 Armillary sphere (1687) showing (from inside out):
earth, celestial sphere, ecliptic, and the horizon

The ancient Greeks became familiar with Babylonian astronomy around 4th century B.C. Eudoxus (408–355 B.C.) developed the "two-sphere model" for astronomy. In this model the stars are considered to be on the celestial sphere (which rotates one revolution a day westward about its pole, the north star) and the sun is on the sphere of the ecliptic, whose equator is the path of the sun and which is inclined to the equator of the celestial sphere at an angle that was about 24° in Eudoxus' time and is about 23½° now. The sphere of the ecliptic is considered to be attached to the celestial sphere and has an apparent rotation eastward of one revolution in a year. Both of these spheres appear to rotate about their poles. See Figure 2.1.

Autolycus, in *On the Rotating Spheres* (333–300 B.C.), introduced a third sphere whose pole is the point directly overhead a particular observer and whose equator is the visible horizon. Thus the angle between the horizon and the celestial equator is equal to the angle (measure at the center of the earth) between the observer and the north pole. Autolycus showed that, for a particular observer, some points (stars) of the celestial sphere are "always visible," some are "always invisible," and some "rise and set." See Figure 2.1.

The earliest known mathematical works that mention spherical geometry are Autolycus' book just mentioned and Euclid's *Phaenomena* [**AT:** Berggren] (300 B.C.). Both of these books use theorems from spherical geometry to solve astrological problems such as *What is the length of daylight on a particular date at a particular latitude?* Euclid used throughout definitions and propositions from spherical geometry. The definitions include *A great circle is the intersection of the sphere by a plane through its center* and *The intersection of the sphere by a plane not through the center forms a (small) circle that is parallel to a unique great circle.* The assumed propositions include, for example, *Suppose two circles are parallel to the same great circle C but on opposite sides; then the two circles are equal if and only if they cut off from some other great circle equal arcs on either side of C.* (We will see similar results in Chapter 10.) There are other more complicated results assumed, including one about the comparison of angles in a spherical triangle; see [**AT:** Berggren], page 25. Thus, it is implied by Autolycus' and Euclid's writings that there were previous works on spherical geometry available to their readers.

Hipparchus of Bithynia (190–120 B.C.) took the spherical coordinates of the Babylonians and applied them to the three spheres (celestial, ecliptic, and horizon). The solution to navigational and astrological problems (such as When will a particular star cross my horizon?) necessitated relating the coordinates on one sphere with the coordinates on the other spheres. This change of coordinates necessitates what we now call spherical trigonometry, and it appears that it was this astronomical problem with spherical coordinates that initiated the study of trigonometry. Plane trigonometry, apparently studied systematically first by Hipparchus, seems to have been originally developed in order to help with spherical trigonometry, which we will study in Chapter 20.

The first systematic account of spherical geometry was *Sphaerica* of Theodosius (around 200 B.C.) It consisted of three books of theorems and construction problems. Most of the propositions of *Sphaerica* were extrinsic theorems and constructions about a sphere as it sits with its center in Euclidean 3-space; but there were also propositions formulated in terms of the intrinsic geometry on the surface of a sphere without reference to either its center or 3-space. We will discuss the distinction between intrinsic and extrinsic later in this chapter.

A more advanced treatise on spherical trigonometry was *On the Sphere* by Menelaus (about 100 A.D.) There exist only edited Arabic versions of this work. In the introduction Menelaus defined a spherical triangle as part of a spherical surface bounded by three arcs of great circles, each less than a semicircle; and he defined the angles of these triangles. Menelaus' treatise expounds geometry on the surface of a sphere in a way analogous to Euclid's exposition of plane geometry in his *Elements*.

Ptolemy (100–178 A.D.) worked in Alexandria and wrote a book on geography, *Geographia* (quoted at the beginning of this chapter), and *Mathematiki Syntaxis* (*Mathematical Collections*), which was the result of centuries of knowledge from Babylonian astronomers and Greek geometers. It became the standard Western work on mathematical astronomy for the next 1400 years. The *Mathematiki Syntaxis* in generally known as the *Almagest*, which is a Latin distortion of the book's name in Arabic that was derived from one of its Greek names. The *Almagest* is important because it is the earliest existing work containing a study of spherical trigonometry, including specific functions, inverse functions, and the computational study of continuous phenomena.

More aspects of the history of spherical geometry will appear later in this book in the appropriate places. For more readings (and references to the primary literature) on this history, see [**HI**: Katz], Chapter 4, and [**HI**: Rosenfeld], Chapter 1.

PROBLEM 2.1   WHAT IS STRAIGHT ON A SPHERE?

Drawing on the understandings about straightness you developed in Problem **1.1**, this problem asks you to investigate the notion of straightness on a sphere. It is important for you to realize that, if you are not building a notion of straightness for yourself (for example, if you are taking ideas from books without thinking deeply about them), then you will have difficulty building a concept of straightness on surfaces other than a plane. Only by developing a personal meaning of straightness for yourself does it become part of your active intuition. We say *active* intuition to emphasize that intuition is in a process of constant change and enrichment, that it is not static.

    a. *Imagine yourself to be a bug crawling around on a sphere. (This bug can neither fly nor burrow into the sphere.) The bug's universe is just the surface; it never leaves it. What is "straight" for this bug? What will the bug see or experience as straight? How can you convince yourself of this? Use the properties of straightness (such as symmetries) that you talked about in Problem 1.1.*

    b. *Show (that is, convince yourself, and give an argument to convince others) that the great circles on a sphere are straight with respect to the sphere, and that no other circles on the sphere are straight with respect to the sphere.*

SUGGESTIONS

Great circles are those circles that are the intersection of the sphere with a plane through the center of the sphere. Examples include longitude lines and the equator on the earth. Any pair of opposite points can be considered as the poles, and thus the equator and longitudes with respect to any pair of opposite points will be great circles. See Figure 2.2.

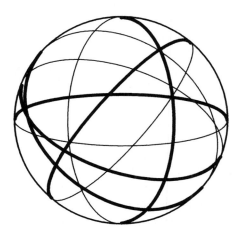

Figure 2.2   Great circles

The first step to understanding this problem is to convince yourself that great circles are straight lines on a sphere. Think what it is about the great circles that would make the bug experience them as straight. To better visualize what is happening on a sphere (or any other surface, for that matter), **you must use models**. This is a point we cannot stress enough. The use of models will become increasingly important in later problems, especially those involving more than one line. You must make lines on a sphere to fully understand what is straight and why. An orange or an old, worn tennis ball work well as spheres, and rubber bands make good lines. Also, you can use ribbon or strips of paper. Try placing these items on the sphere along different curves to see what happens.

Also look at the symmetries from Problem **1.1** to see if they hold for straight lines on the sphere. The important thing here is to **think in terms of the surface of the sphere, not the solid 3-dimensional ball**. Always try to imagine how things would look from the bug's point of view. A good example of how this type of thinking works is to look at an insect called a water strider. The water strider walks on the surface of a pond and has a very 2-dimensional perception of the world around it — to the water strider, there is no up or down; its whole world consists of the 2-dimensional plane of the water. The water strider is very sensitive to motion and vibration on the water's surface, but it can be approached from above or below without its knowledge. Hungry birds and fish take advantage of this fact. This is the type of thinking needed to visualize adequately properties of straight lines on the sphere. For more discussion of water striders and other animals with their own varieties of intrinsic observations, see the delightful book *The View from the Oak*, by Judith and Herbert Kohl [**NA**: Kohl and Kohl].

Definition. Paths that are intrinsically straight on a sphere (or other surfaces) are called *geodesics*.

This leads us to consider the concept of intrinsic or geodesic curvature versus extrinsic curvature. As an outside observer looking at the sphere in 3-space, all paths on the sphere, even the great circles, are curved — that is, they exhibit *extrinsic curvature*. But relative to the surface of the sphere (*intrinsically*), the lines may be straight and thus have intrinsic curvature zero. See the last section of this chapter, Intrinsic Curvature. Be sure to understand this difference and to see why all symmetries (such as reflections) must be carried out intrinsically, or from the bug's point of view.

It is natural for you to have some difficulty experiencing straight-ness on surfaces other than the 2-dimensional plane; it is likely that you will start to look at spheres and the curves on spheres as 3-dimensional objects. Imagining that you are a 2-dimensional bug walking on a sphere helps you to shed your limiting extrinsic 3-dimensional vision of the curves on a sphere and to experience straightness intrinsically. Ask yourself the following:

- What does the bug have to do, when walking on a non-planar surface, in order to walk in a straight line?
- How can the bug check if it is going straight?

Experimentation with models plays an important role here. Working with models that *you create* helps you to experience great circles as, in fact, the only straight lines on the surface of a sphere. Convincing yourself of this notion will involve recognizing that straightness on the plane and straightness on a sphere have common elements. When you are comfortable with "great-circle-straightness," you will be ready to transfer the symmetries of straight lines on the plane to great circles on a sphere and, later, to geodesics on other surfaces. Here are some activities that you can try, or visualize, to help you experience great circles and their intrinsic straightness on a sphere. However, it is better for you to come up with your own experiences.

- Stretch something elastic on a sphere. It will stay in place on a great circle, but it will not stay on a small circle if the sphere is slippery. Here, the elastic follows a path that is approximately the shortest because a stretched elastic always moves so that it will be shorter. This a very useful practical criterion of straightness.

- Roll a ball on a straight chalk line (or straight on a freshly painted floor!). The chalk (or paint) will mark the line of contact on the sphere, and it will form a great circle.

- Take a narrow stiff ribbon or strip of paper that does not stretch, and lay it "flat" on a sphere. It will only lie (without folds and creases) along a great circle. Do you see how this property is related to local symmetry? This is sometimes called the *Ribbon Test*. (For further discussion of the Ribbon Test, see Problems **3.4** and **7.6** of [**DG:** Henderson].)

◆ The feeling of turning and "nonturning" comes up. Why is it that on a great circle there is no turning and on a latitude line there is turning? Physically, in order to avoid turning, the bug has to move its left feet the same distance as its right feet. On a non-great circle (for example, a latitude line that is not the equator), the bug has to walk faster with the legs that are on the side closer to the equator. This same idea can be experienced by taking a small toy car with its wheels fixed to parallel axes so that, on a plane, it rolls along a straight line. On a sphere, the car will roll around a great circle; but it will not roll around other curves.

◆ Also notice that, on a sphere, straight lines are intrinsic circles (points on the surface a fixed distance along the surface away from a given point on the surface) — special circles whose circumferences are straight! Note that the equator is a circle with two intrinsic centers: the north pole and the south pole. In fact, any circle (such as a latitude circle) on a sphere has two intrinsic centers.

These activities will provide you with an opportunity to investigate the relationships between a sphere and the geodesics of that sphere. Along the way, your experiences should help you to discover how great circles on a sphere have most of the same symmetries as straight lines on a plane.

You should pause and not read further until you have expressed your thinking and ideas about this problem.

SYMMETRIES OF GREAT CIRCLES

Reflection-through-itself symmetry: We can see this globally by placing a hemisphere on a flat mirror. The hemisphere together with the image in the mirror exactly recreates a whole sphere. Figure 2.3 shows a reflection through the great circle *g*.

Reflection-perpendicular-to-itself symmetry: A reflection through any great circle will take any great circle (for example, *g*′ in Figure 2.3) perpendicular to the original great circle onto itself.

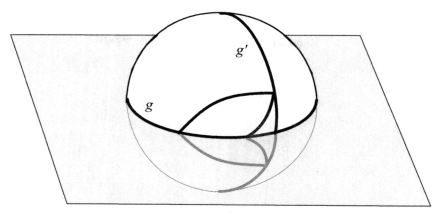

Figure 2.3   Reflection-through-itself symmetry

Half-turn symmetry: A rotation through half of a full revolution about any point P on a great circle interchanges the part of the great circle on one side of P with the part on the other side of P. See Figure 2.4.

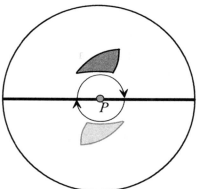

Figure 2.4   Half-turn symmetry

Rigid-motion-along-itself symmetry: For great circles on a sphere, we call this a translation along the great circle or a rotation around the poles of that great circle. This property of being able to move rigidly along itself is not unique to great circles because any circle on the sphere will also have the same symmetry. See Figure 2.5.

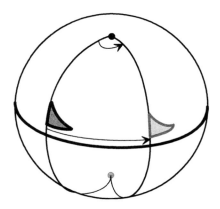

Figure 2.5   Rigid-motion-along-itself symmetry

Central symmetry, or point symmetry: Viewed intrinsically (from the 2-dimensional bug's point-of-view), central symmetry through a point P on the sphere sends any point A to the point at the same great circle distance from P but on the opposite side. See Figure 2.6.

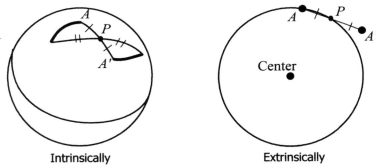

Intrinsically   Extrinsically
Figure 2.6   Central symmetry through P

Extrinsically (viewing the sphere in 3-space) central symmetry through P would send A to a point off the surface of the sphere as shown in Figure 2.6. The only extrinsic central symmetry of the sphere (and the only one for great circles on the sphere) is through the center of the sphere (which is not *on* the sphere). The transformation that is intrinsically central symmetry is extrinsically half-turn symmetry (about the diameter through P). Intrinsically, as on a plane, central symmetry does not differ from half-turn symmetry with respect to the end result. This distinction between intrinsic and extrinsic is important to experience at this point.

3-dimensional-rotation symmetry: This symmetry does not hold for great circles in 3-space; however, it does hold for great circles in a 3-sphere. See Problem **22.5**.

You will probably notice that other objects on the sphere, besides great circles, have some of the symmetries mentioned here. It is important for you to construct such examples. This will help you to realize that straightness and the symmetries discussed here are intimately related.

*EVERY GEODESIC IS A GREAT CIRCLE

Notice that you were not asked to prove that *every geodesic* (intrinsic straight line) *on the sphere is a great circle*. This is true but more difficult to prove. Many texts simply *define* the great circles to be the "straight lines" (geodesics) on the sphere. We have not taken that approach. We have shown that the great circles are intrinscially straight (geodesics), and it is clear that two points on the sphere are always joined by a great circle arc, which shows that there are sufficient great-circle geodesics to do the geometry we wish.

To show that great circles are the only geodesics involves some notions from differential geometry. In Problem **3.2b** of [**DG**: Henderson] this is proved using special properties of plane curves. More generally, a geodesic satisfies a differential equation with the initial condition being a point on the geodesic and the direction of the geodesic at that point (see Problem **8.4b** of [**DG**: Henderson]). Thus it follows from the analysis theorem on the *existence and uniqueness of solutions to differential equations* that

> **THEOREM 2.1.** *At every point and in every direction on a smooth surface there is a unique geodesic going from that point in that direction.*

From this it follows that all geodesics on a sphere are great circles. *Do you see why?*

*INTRINSIC CURVATURE

You have tried wrapping the sphere with a ribbon and noticed that the ribbon will only lie flat along a great circle. (If you haven't experienced this yet, then do it now before you go on.) Arcs of great circles are the only paths on a sphere's surface that are tangent to a straight line on a piece of paper wrapped around the sphere.

If you wrap a piece of paper tangent to the sphere around a latitude circle (see Figure 2.7), then, extrinsically, the paper will form a portion of a cone and the curve on the paper will be an arc of a circle when the paper is flattened. The *intrinsic curvature* of a path on the surface of a sphere can be defined as the curvature (1/radius) that one gets when one "unwraps" the path onto a plane. For more details, see Chapter 3 of [**DG**: Henderson].

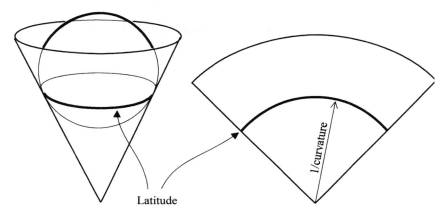

Latitude

Figure 2.7   Finding the intrinsic curvature

Differential geometers often talk about intrinsically straight paths (geodesics) in terms of the velocity vector of the motion as one travels at a constant speed along that path. (The velocity vector is tangent to the curve along which the bug walks.) For example, as you walk along a great circle, the velocity vector to the circle changes direction, extrinsically, in 3-space where the change in direction is toward the center of the sphere. "Toward the center" is not a direction that makes sense to a 2-dimensional bug whose whole universe is the surface of the sphere. Thus, the bug does not experience the velocity vectors as changing direction at points along the great circle; however, along non-great circles the velocity vector will be experienced as changing in the direction of the closest center of the circle. In differential geometry, the rate of change, from the bug's point of view, is called the *covariant* (or *intrinsic*) *derivative*. As the bug traverses a geodesic, the covariant derivative of the velocity vector is zero. This can also be expressed in terms of *parallel transport*, which is discussed in Chapters 7, 8, and 10 of this text. See [**DG**: Henderson] for discussions of these ideas in differential geometry.

Chapter 3

WHAT IS AN ANGLE?

> A (plane) *angle* is the inclination to one another of two lines in a plane which meet one another and do not lie in a straight line. — Euclid, *Elements*, Definition 8 [Appendix A]

In this chapter you will be thinking about angles. In **3.1** we will investigate various notions and definitions of angles and what it means for them to be considered to be the same (*congruent*). In **3.2** we will prove the important *Vertical Angle Theorem* (VAT). It is not necessary to do these parts in order — you may find it easier to do Problem **3.2** before **3.1** because it may help you think about angles. In a sense, you should be working on **3.1** and **3.2** at the same time because they are so closely intertwined. This provides a valuable opportunity to apply and reflect on what you have learned about straightness in Chapters 1 and 2. This will also be helpful in the further study of straightness in Chapters 4 and 5; but, if you wish, you may study this chapter after Chapters 4 and 5.

PROBLEM 3.1 WHAT IS AN ANGLE?

> *Give some possible definitions of the term "angle." Do all of these definitions apply to the plane as well as to spheres? What are the advantages and disadvantages of each?*
>
> *For each definition, what does it mean for two angles to be congruent? How can we check?*

SUGGESTIONS

Etymologically, "angle" comes through Old English, Old French, Old German, Latin, and Greek words for "hook." Textbooks usually give some variant of the definition: *An angle is the union of two rays (or segments) with a common endpoint.*

If we start with two straight line segments with a common endpoint and then add squiggly parts onto the ends of each one, would we say that the angle has changed as a result? Likewise, look at the angle formed at the lower-left-hand corner of this piece of paper. Even first grade students will recognize this as an example of an angle. Now, tear off the corner (at least in your imagination). Is the angle still there, on the piece you tore off? Now tear away more of the sides of the angles, being careful not to tear through the corner. The angle is still there at the corner, isn't it? See Figure 3.1.

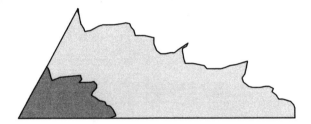

Figure 3.1   Where is the angle?

So what part of the angle determines how large the angle is, or if it is an angle at all? What is the angle? It seems it cannot be merely a union of two rays. Here is one of the many cases where children seem to know more than we do. Paying attention to these insights, can we get better definitions of "angle"? Do not expect to find one formal definition that is completely satisfactory; it seems likely that no formal definition can capture all aspects of our experience of what an angle is.

There are at least three different perspectives from which we can define "angle," as follows:

- a *dynamic* notion of angle — angle as movement;

- angle as *measure*; and,

- angle as *geometric shape*.

A *dynamic* notion of angle involves an action: a rotation, a turning point, or a change in direction between two lines. Angle as *measure* may be thought of as the length of a circular arcs or the ratio between areas of circular sectors. Thought of as a ***geometric shape,*** an angle may be seen as the delineation of space by two intersecting lines. Each of these perspectives carries with it methods for checking angle congruency. You can check the congruency of two dynamic angles by verifying that the actions involved in creating or replicating them are the same. If you feel that an angle is a measure, then you must verify that both angles have the same measure. If you describe angles as geometric shapes, then you describe how one angle can be made to coincide with the other using isometries. Which of the above definitions has the most meaning for you? Are there any other useful ways of describing angles?

Note that we sometimes talk about ***directed angles,*** or angles with direction. When considered as directed angles, we say that the angles α and β in Figure 3.2 are not the same but have equal magnitude and opposite directions (or sense). Note the similarity to the relationship between line segments and vectors.

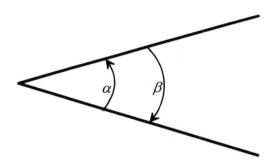

Figure 3.2   Directed angles

PROBLEM 3.2 VERTICAL ANGLE THEOREM (VAT)

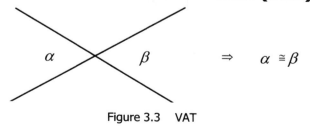

Figure 3.3   VAT

*Prove: Opposite angles formed by two intersecting straight lines are congruent. [Note: Angles such as α and β are called **vertical*** *angles*.] *What properties of straight lines and/or the plane are you using in your proof? Does your proof also work on a sphere? Why? Which definitions from Problem* **3.2** *are you using in your proof?*

Show how you would "move" α to make it coincide with β. We do not have in mind a formal two-column proof that used to be in American high school geometry. Mathematicians in actual practice usually use "proof" to mean "a convincing communication that answers — Why?" This is the notion of proof we ask you to use. There are three features of a proof:

- It must *communicate* (the words and drawings need to clearly express what it is that you want to say — and they must be understandable to your reader and/or listener.)

- It must be *convincing* (to yourself, to your fellow students, and to your teacher; preferably it should be convincing to someone who was originally skeptical).

- It must *answer — Why?* (Why is it true? What does it mean? Where did it come from?)

The goal is understanding. Without understanding we will never be fully satisfied. With understanding we want to expand that understanding and to communicate it to others.

Symmetries were an important element of your solutions for Problems **1.1** and **2.1**. They will be very useful for this problem as well. It is perfectly valid to think about measuring angles in this problem, but proofs utilizing line symmetries are generally simpler. It often helps to think of the vertical angles as whole geometric figures. Also, keep in mind that there are many different ways of looking at angles, so there are many ways of proving the vertical angle theorem. Make sure that your notions of angle and angle congruency in Problem **3.1** are consistent with your proofs in Problem **3.2**, and vice versa. Any of the definitions from **3.1** can, separately or together, help you prove the Vertical Angle Theorem.

You should pause and not read further until you have expressed your own thinking and ideas about Problems 3.1 and 3.2.

HINTS FOR THREE DIFFERENT PROOFS

In the following section, we will give hints for three different proofs of the Vertical Angle Theorem. Note that a particular notion of angle is assumed in each proof. Pick one of the proofs, or find your own different proof that is consistent with a notion of angle and angle congruence that is most meaningful to you.

1st proof:

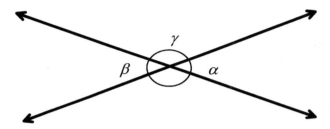

Figure 3.4   VAT using angle as measure

Each line creates a 180° angle. Thus, $\alpha + \gamma = \beta + \gamma$. See Figure 3.4.

Therefore, we can conclude that $\alpha \cong \beta$. But why is this so? Is it always true that if we subtract a given angle from two 180° angles then the remaining angles are congruent? See Figure 3.5.

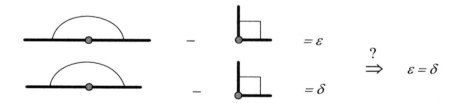

Figure 3.5   Subtracting angles and measures

Numerically, it does not make any difference how we subtract an angle, but geometrically it makes a big difference. Behold Figure 3.6! Here, ε really cannot be considered the same as δ. Thus, measure does not completely express what we *see* in the geometry of this situation. If you wish to salvage this notion of angle as measure, then you must explain *why* it is that in this proof of the Vertical Angle Theorem γ can be subtracted from both sides of the equation $\alpha + \gamma = \beta + \gamma$.

Figure 3.6 δ is not the same as ε

2nd proof: Consider two overlapping lines and choose any point on them. Rotate one of the lines, maintaining the point of intersection and making sure that the other line remains fixed. See Figure 3.7.

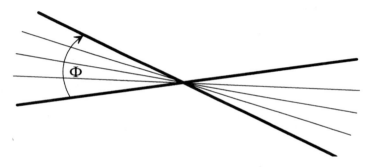

Figure 3.7   VAT using angle as rotation

What happens? What notion of angle and angle congruency is at work here?

3rd proof: What symmetries will take α onto β? See Figure 3.3 or 3.4. Use the properties of straight lines you investigated in Chapters 1 and 2.

Chapter 4

STRAIGHTNESS ON CYLINDERS AND CONES

> If a cut were made through a cone parallel to its base, how should we conceive of the two opposing surfaces which the cut has produced — as equal or as unequal? If they are unequal, that would imply that a cone is composed of many breaks and protrusions like steps. On the other hand if they are equal, that would imply that two adjacent intersection planes are equal, which would mean that the cone, being made up of equal rather than unequal circles, must have the same appearance as a cylinder; which is utterly absurd.
>
> — Democritus of Abdera (~460 – ~380 B.C.)

This quote shows that cylinders and cones were the subject of mathematical inquiry before Euclid (~365 – ~300 B.C.). In this chapter we investigate straightness on cones and cylinders. You should be comfortable with straightness as a *local intrinsic notion* — this is the bug's view. This notion of straightness is also the basis for the notion of *geodesics* in differential geometry. Chapters 4 and 5 can be covered in either order, but we think that the experience with cylinders and cones in **4.1** will help the reader to understand the hyperbolic plane in **5.1**. If the reader is comfortable with straightness as a local intrinsic notion, then it is also possible to skip Chapter 4 if Chapters 18 and 24 on geometric manifolds are not going to be covered. However, we suggest that you read the sections at the end of this chapter — Is "Shortest" Always "Straight"? and Relations to Differential Geometry — at least enough to find out what Euclid's Fourth Postulate (see Appendix A) has to do with cones and cylinders.

When looking at great circles on the surface of a sphere, we were able (except in the case of central symmetry) to see all the symmetries of straight lines from global extrinsic points of view. For example, a great circle extrinsically divides a sphere into two hemispheres that are mirror images of each other. Thus on a sphere, it is a natural tendency to use the more familiar and comfortable extrinsic lens instead of taking the bug's local and intrinsic point of view. However, on a cone and cylinder you must use the local, intrinsic point of view because there is no extrinsic view that will work except in special cases.

PROBLEM 4.1   STRAIGHTNESS ON CYLINDERS AND CONES

a.   *What lines are straight with respect to the surface of a cylinder or a cone? Why? Why not?*

b.   *Examine*:

- ◆ *Can geodesics intersect themselves on cylinders and cones?*

- ◆ *Can there be more than one geodesic joining two points on cylinders and cones?*

- ◆ *What happens on cones with varying cone angles, including cone angles greater than 360°? These are discussed starting in the next section.*

SUGGESTIONS

Problem **4.1** is similar to Problem **2.1**, but this time the surfaces are cylinders and cones.

Make paper models, but consider the cone or cylinder as continuing indefinitely with no top or bottom (except, of course, at the cone point). Again, imagine yourself as a bug whose whole universe is a cone or cylinder. As the bug crawls around on one of these surfaces, what will the bug experience as straight? As before, paths that are straight with respect to a surface are often called the "geodesics" for the surface.

As you begin to explore these questions, it is likely that many other related geometric ideas will arise. Do not let seemingly irrelevant excess geometric baggage worry you. Often, you will find yourself getting lost in a tangential idea, and that's understandable. Ultimately, however, the exploration of related ideas will give you a richer understanding of the scope and depth of the problem. In order to work through possible confusion on this problem, try some of the following suggestions others have found helpful. Each suggestion involves constructing or using models of cones and cylinders.

◆ You may find it helpful to explore cylinders first before beginning to explore cones. This problem has many aspects, but focusing at first on the cylinder will simplify some things.

◆ If we make a cone or cylinder by rolling up a sheet of paper, will "straight" stay the same for the bug when we unroll it? Conversely, if we have a straight line drawn on a sheet of paper and roll it up, will it continue to be experienced as straight for the bug crawling on the paper? *We are assuming here that the paper will not stretch and its thickness is negligible.*

◆ Lay a stiff ribbon or straight strip of paper on a cylinder or cone. Convince yourself that it will follow a straight line with respect to the surface. Also, convince yourself that straight lines on the cylinder or cone, *when looked at locally and intrinsically*, have the same symmetries as on the plane.

◆ If you intersect a cylinder by a flat plane and unroll it, what kind of curve do you get? Is it ever straight? (One way to see this curve is to dip a paper cylinder into water.)

◆ On a cylinder or cone, can a geodesic ever intersect itself? How many times? This question is explored in more detail in Problem **4.2**, which the interested reader may turn to now.

◆ Can there be more than one geodesic joining two points on a cylinder or cone? How many? Is there always at least one? Again this question is explored in more detail in Problem **4.2**.

There are several important things to keep in mind while working on this problem. First, **you absolutely must make models**. If you attempt to visualize lines on a cone or cylinder, you are bound to make claims that you would easily see are mistaken if you investigated them on an actual cone or cylinder. Many students find it helpful to make models using transparencies.

Second, as with the sphere, you must think about lines and triangles on the cone and cylinder in an intrinsic way — always looking at things from a bug's point of view. We are not interested in what's happening in 3-space, only what you would see and experience if you were restricted to the surface of a cone or cylinder.

And last, but certainly not least, you must look at cones of different shapes, that is, cones with varying cone angles.

CONES WITH VARYING CONE ANGLES

Geodesics behave differently on differently shaped cones. So an important variable is the cone angle. The ***cone angle*** is generally defined as the angle measured around the point of the cone on the surface. Notice that this is an intrinsic description of angle. The bug could measure a cone angle (in radians) by determining the circumference of an intrinsic circle with center at the cone point and then dividing that circumference by the radius of the circle. We can determine the cone angle extrinsically in the following way: Cut the cone along a ***generator*** (a line on the cone through the cone point) and flatten the cone. The measure of the cone angle is then the angle measure of the flattened planar sector.

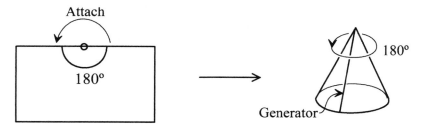

Figure 4.1   Making a 180° cone

For example, if we take a piece of paper and bend it so that half of one side meets up with the other half of the same side, we will have a 180-degree cone (Figure 4.1). A 90° cone is also easy to make — just use the corner of a paper sheet and bring one side around to meet the adjacent side.

Also be sure to look at larger cones. One convenient way to do this is to make a cone with a variable cone angle. This can be accomplished by taking a sheet of paper and cutting (or tearing) a slit from one edge to the center. (See Figure 4.2.) A rectangular sheet will work but a circular sheet is easier to picture. Note that it is not necessary that the slit be straight!

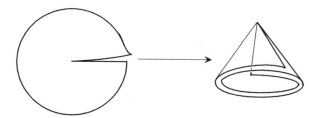

Figure 4.2   A cone with variable cone angle (0–360°)

You have already looked at a 360° cone — it's just a plane. The cone angle can also be larger than 360°. A common larger cone is the 450° cone. You probably have a cone like this somewhere on the walls, floor, and ceiling of your room. You can easily make one by cutting a slit in a piece of paper and inserting a 90° slice (360° + 90° = 450°) as in Figure 4.3.

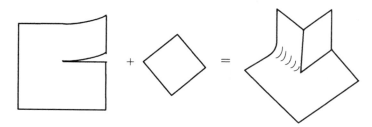

Figure 4.3   How to make a 450° cone

You may have trouble believing that this is a cone, but remember that just because it cannot hold ice cream does not mean it is not a cone. If the folds and creases bother you, they can be taken out — the cone will look ruffled instead. It is important to realize that when you change the shape of the cone like this (that is, by ruffling), you are only changing its extrinsic appearance. Intrinsically (from the bug's point of view) there is no difference. You can even ruffle the cone so that it will hold ice cream if you like, although changing the extrinsic shape in this way is not useful to a study of its intrinsic behavior.

It may be helpful for you to discuss some definitions of a cone, such as the following: *Take any simple (non-intersecting) closed curve **a** on a* *sphere and the center **P** of the sphere. A **cone** is the union of the rays that start at **P** and go through each point on **a**.* The cone angle is then equal to (length of **a**)/(radius of sphere), in radians. Do you see why?

You can also make a cone with variable angle of more than 180°: Take two sheets of paper and slit them together to their centers as in Figure 4.4. Tape the right side of the top slit to the left side of the bottom slit as pictured. Now slide the other sides of the slits. Try it!

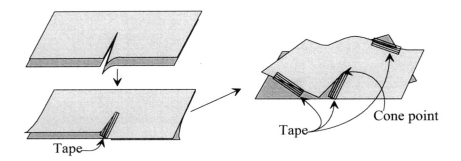

Figure 4.4   Variable cone angle larger than 360°

Experiment by making paper examples of cones like those shown in Figure 4.4. What happens to the triangles and lines on a 450° cone? Is the shortest path always straight? Does every pair of points determine a straight line?

Finally, also consider line symmetries on the cone and cylinder. Check to see if the symmetries you found on the plane will work on these surfaces, and remember to think intrinsically and locally. A special class of geodesics on the cone and cylinder are the generators. These are the straight lines that go through the cone point on the cone or go parallel to the axis of the cylinder. These lines have some extrinsic symmetries (*can you see which ones*?), but in general, geodesics have only local, intrinsic symmetries. Also, on the cone, think about the region near the cone point — what is happening there that makes it different from the rest of the cone?

It is best if you experiment with paper models to find out what geodesics look like on the cone and cylinder before reading further.

GEODESICS ON CYLINDERS

Let us first look at the three classes of straight lines on a cylinder. When walking on the surface of a cylinder, a bug might walk along a vertical generator. See Figure 4.5.

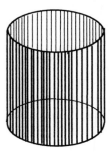

Figure 4.5   Vertical generators are straight

It might walk along an intersection of a horizontal plane with the cylinder, what we will call a *great circle*. See Figure 4.6

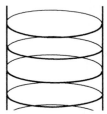

Figure 4.6   Great circles are intrinsically straight

Or, the bug might walk along a spiral or helix of constant slope around the cylinder. See Figure 4.7.

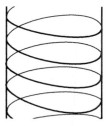

Figure 4.7   Helixes are intrinsically straight

Why are these geodesics? How can you convince yourself? And why are these the only geodesics?

GEODESICS ON CONES

Now let us look at the classes of straight lines on a cone.

Walking along a generator: When looking at straight paths on a cone, you will be forced to consider straightness at the cone point. You might decide that there is no way the bug can go straight once it reaches the cone point, and thus a straight path leading up to the cone point ends there. Or you might decide that the bug can find a continuing path that has at least some of the symmetries of a straight line. Do you see which path this is? Or you might decide that the straight continuing path(s?) is the limit of geodesics that just miss the cone point. See Figure 4.8.

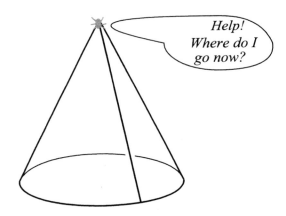

Help! Where do I go now?

Figure 4.8   Bug walking straight over the cone point

Walking straight and around: If you use a ribbon on a 90° cone, then you can see that this cone has a geodesic like the one depicted in Figure 4.9. This particular geodesic intersects itself. However, check to see that this property depends on the cone angle. In particular, if the cone angle is more than 180°, then geodesics do not intersect themselves. And if the cone angle is less than 90°, then geodesics (except for generators) intersect at least two times. Try it out! Later, in Problem **4.2**, we will describe a tool that will help you determine how the number of self-intersections depends on the cone angle.

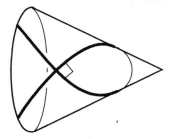

Figure 4.9   A geodesic intersecting itself on a 90° cone

*PROBLEM 4.2   GLOBAL PROPERTIES OF GEODESICS

Now we will look more closely at long geodesics that wrap around on a cylinder or cone. Several questions have arisen.

    a. *How do we determine the different geodesics connecting two points? How many are there? How does it depend on the cone angle? Is there always at least one geodesic joining each pair of points? How can we justify our conjectures?*

    b. *How many times can a geodesic on a cylinder or cone intersect itself? How are the self-intersections related to the cone angle? At what angle does the geodesic intersect itself? How can we justify these relationships?*

SUGGESTIONS

Here we offer the tool of covering spaces, which may help you explore these questions. The method of coverings is so named because it utilizes layers (or sheets) that each cover the surface. We will first start with a cylinder because it is easier and then move on to a cone.

*n-SHEETED COVERINGS OF A CYLINDER

To understand how the method of coverings works, imagine taking a paper cylinder and cutting it axially (along a vertical generator) so that it unrolls into a plane. This is probably the way you constructed cylinders to study this problem before. The unrolled sheet (a portion of the plane) is said to be a *1-sheeted covering* of the cylinder. See Figure 4.10. If you marked two points on the cylinder, *A* and *B*, as indicated in the figure, when the cylinder is cut and unrolled into the covering, these two points become two points on the covering (which are labeled by the same letters in the figure). The two points on the covering are said to be *lifts* of the points on the cylinder.

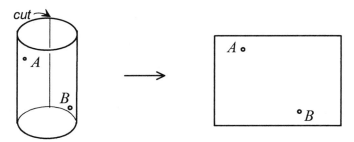

Figure 4.10   A 1-sheeted covering of a cylinder

Now imagine attaching several of these "sheets" together, end to end. When rolled up, each sheet will go around the cylinder exactly once — they will each cover the cylinder. (Rolls of toilet paper or paper towels give a rough idea of coverings of a cylinder.) Also, each sheet of the covering will have the points *A* and *B* in identical locations. You can see this (assuming the paper thickness is negligible) by rolling up the coverings and making points by sticking a sharp object through the cylinder. This means that all the *A*'s are coverings of the same point on the cylinder and all the *B*'s are coverings of the same point on the cylinder. We just have on the covering several representations, or *lifts*, of each point on the cylinder. Figure 4.11 depicts a 3-sheeted covering space for a cylinder and six geodesics joining *A* to *B*. (One of them is the most direct path from *A* to *B* and the others spiral once, twice, or three times around the cylinder in one of two directions.)

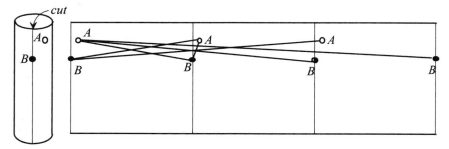

Figure 4.11   A 3-sheeted covering space for a cylinder

We could also have added more sheets to the covering on either the right or left side. You can now roll these sheets back into a cylinder and see what the geodesics look like. Remember to roll sheets up so that each sheet of the covering covers the cylinder exactly once — all of the vertical lines between the coverings should lie on the same generator of the cylinder. Note that if you do this with ordinary paper, part or all of some geodesics will be hidden, even though they are all there. It may be easier to see what's happening if you use transparencies.

This method works because straightness is a local intrinsic property. Thus, lines that are straight when the coverings are laid out in a plane will still be straight when rolled into a cylinder. Remember that bending the paper does not change the intrinsic nature of the surface. Bending only changes the curvature that we see extrinsically. It is important always to look at the geodesics from the bug's point of view. The cylinder and its covering are locally isometric.

Use coverings to investigate Problem **4.2** on the cylinder. The global behavior of straight lines may be easier to see on the covering.

*n-SHEETED (BRANCHED) COVERINGS OF A CONE

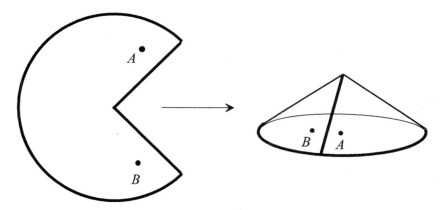

Figure 4.12   1-sheeted covering of a 270° cone

Figure 4.12 shows a 1-sheeted covering of a cone. The sheet of paper and the cone are locally isometric except at the cone point. The cone point is called a **branch point** of the covering. We talk about lifts of points on the cone in the same way as on the cylinder. In Figure 4.12 we depict a 1-sheeted covering of a 270° cone and label two points and their lifts.

A 4-sheeted covering space for a cone is depicted in Figure 4.13. Each of the rays drawn from the center of the covering is a lift of a single ray on the cone. Similarly, the points marked on the covering are the lifts of the points *A* and *B* on the cone. In the covering there are four segments joining a lift of *A* to different lifts of *B*. Each of these segments is the lift of a different geodesic segment joining *A* to *B*.

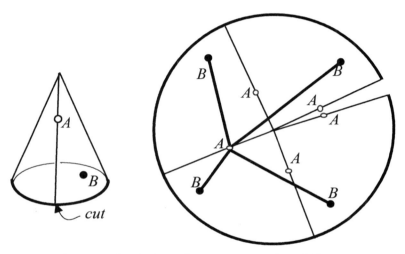

Figure 4.13   4-sheeted covering space for a 89° cone

Think about ways that the bug can use coverings as a tool to expand its exploration of surface geodesics. Also, think about ways you can use coverings to justify your observations in an intrinsic way. It is important to be precise; you don't want the bug to get lost! Count the number of ways in which you can connect two points with a straight line and relate those countings with the cone angle. Does the number of straight paths only depend on the cone angle? Look at the 450° cone and see if it is always possible to connect any two points with a straight line. **Make paper models!** It is not possible to get an equation that relates the cone angle to the number of geodesics joining every pair of points. However, it is possible to find a formula that works for most pairs. Make covering spaces for cones of different size angles and refine the guesses you have already made about the numbers of self-intersections.

In studying the self-intersections of a geodesic *l* on a cone, it may be helpful for you to consider the ray ***R*** that is perpendicular to the line *l*. (See Figure 4.14.) Now study one lift of the geodesic *l* and its relationship to the lifts of the ray ***R***. Note that the seams between individual wedges are lifts of ***R***.

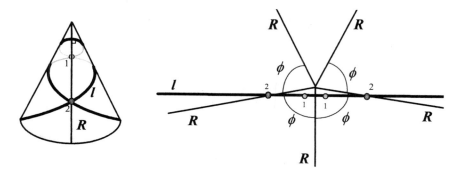

Figure 4.14   Self-intersections on a cone with angle ϕ

LOCALLY ISOMETRIC

By now you should realize that when a piece of paper is rolled or bent into a cylinder or cone, the bug's local and intrinsic experience of the surface does not change except at the cone point. Extrinsically, the piece of paper and the cone are different, but in terms of the local geometry intrinsic to the surface they differ only at the cone point.

Two geometric spaces, **G** and **H**, are said to be ***locally isometric*** at the points *G* in **G** and *H* in **H** if the local intrinsic experience at *G* is the same as the experience at *H*. That is, there are neighborhoods of *G* and *H* that are identical in terms of their intrinsic geometric properties. A cylinder and the plane are locally isometric (at each point) and the plane and a cone are locally isometric except at the cone point. Two cones are locally isometric at the cone points only if their cone angles are the same.

Because cones and cylinders are locally isometric with the plane, locally they have the same geometric properties. Later, we will show that a sphere is not locally isometric with the plane — *be on the lookout for a result that will imply this.*

IS "SHORTEST" ALWAYS "STRAIGHT"?

We are often told that "a straight line is the shortest distance between two points," but is this really true? As we have already seen on a sphere, two points not opposite each other are connected by two straight paths (one going one way around a great circle and one going the other way). Only one of these paths is shortest. The other is also straight, but not the shortest straight path.

Consider a model of a cone with angle 450°. Notice that such cones appear commonly in buildings as so-called "outside corners" (see Figure 4.3). It is best, however, to have a paper model that can be flattened.

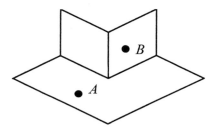

Figure 4.15   There is no straight (symmetric) path from *A* to *B*

Use your model to investigate which points on the cone can be joined by straight lines (in the sense of having reflection-in-the-line symmetry). In particular, look at points such as those labeled *A* and *B* in Figure 4.15. Convince yourself that there is no path from *A* to *B* that is straight (in the sense of having reflection-in-the-line symmetry), and for these points the shortest path goes through the cone point and thus is not straight (in the sense of having symmetry).

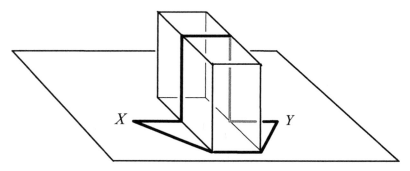

Figure 4.16   The shortest path is not straight (in the sense of symmetry)

Here is another example: Think of a bug crawling on a plane with a tall box sitting on that plane (refer to Figure 4.16). This combination surface — the plane with the box sticking out of it — has eight cone points. The four at the top of the box have 270° cone angles, and the four at the bottom of the box have 450° cone angles (180° on the box and 270° on the plane). What is the shortest path between points *X* and *Y*, points on opposite sides of the box? Is the straight path the shortest? Is the shortest path straight? To check that the shortest path is not straight, try to see that at the bottom corners of the box the two sides of the path have different angular measures. (In particular, if *X* and *Y* are close to the box, then the angle on the box side of the path measures a little more than 180° and the angle on the other side measures almost 270°.)

*RELATIONS TO DIFFERENTIAL GEOMETRY

We see that sometimes a straight path is not shortest and the shortest path is not straight. Does it then makes sense to say (as most books do) that in Euclidean geometry a straight line is the shortest distance between two points? In differential geometry, on "smooth" surfaces, "straight" and "shortest" are more nearly the same. A *smooth* surface is essentially what it sounds like. More precisely, a surface is smooth at a point if, when you zoom in on the point, the surface becomes indistinguishable from a flat plane. (For details of this definition, see Problem **4.1** in [**DG**: Henderson]. See also the last section and especially the endnote in Chapter 1.) Note that a cone is not smooth at the cone point, but a sphere and a cylinder are both smooth at every point. The following is a theorem from differential geometry:

> **THEOREM 4.1:** *If a surface is smooth, then an intrinsically straight line (geodesic) on the surface is always the shortest path between "nearby" points. If the surface is also complete (every geodesic on it can be extended indefinitely), then any two points can be joined by a geodesic that is the shortest path between them.* See [**DG**: Henderson], Problems **7.4b** and **7.4d**.

Consider a planar surface with a hole removed. Check that for points near opposite sides of the hole, the shortest path (on the planar surface with hole removed) is not straight because the shortest path must go around the hole. *We encourage the reader to discuss how each of the previous examples and problems is in harmony with this theorem.*

Note that the statement "every geodesic on the surface can be extended indefinitely" is a reasonable interpretation of Euclid's Second Postulate: *Every limited straight line can be extended indefinitely to a (unique) straight line* [Appendix A, Postulate 2]. Note that the Second Postulate does not hold on a cone unless you consider geodesics to continue through the cone point.

Also, Euclid defines a right angle as follows: *When a straight line intersects another straight line such that the adjacent angles are equal to one another, then the equal angles are called **right angles*** [Appendix A, Definition 10]. Note that if you consider geodesics to continue through the cone point, then right angles at a cone point are not equal to right angles at points where the cone is locally isometric to the plane.

And Euclid goes on to state as his Fourth Postulate: *All right angles are equal* [Appendix A, Postulate 4]. Thus, Euclid's Second Postulate or Fourth Postulate rules out cones and any surface with isolated cone points. What is further ruled out by Euclid's Fourth Postulate would depend on formulating more precisely just what it says. It is not clear (at least to the authors!) whether there is something we would want to call a surface that could be said to satisfy Euclid's Fourth Postulate and not be a smooth surface. However, it is clear that Euclid's postulate at least gives part of the meaning of "smooth surface," because it rules out isolated cone points.

When we were in high school geometry class we could not understand why Euclid would have made such a postulate as his Postulate 4 — how could they possibly *not* be equal? In this chapter we have discovered that on cones right angles are not all equal.

Chapter 5

STRAIGHTNESS ON HYPERBOLIC PLANES

> [To son János:] For God's sake, please give it [work on hyperbolic geometry] up. Fear it no less than the sensual passion, because it, too, may take up all your time and deprive you of your health, peace of mind and happiness in life.
> — Wolfgang Bolyai (1775–1856)
> [**EM**: Davis and Hersh], page 220

We now study hyperbolic geometry. This chapter may be skipped if the reader will not be covering Chapter 17 and if, in the remainder of this book, the reader leaves out all mentions of hyperbolic planes. However, to skip studying hyperbolic planes would be to skip an important notion in the history of geometry and also to skip the geometry that may be the basis of our physical universe.

As with the cone and cylinder, we must use an intrinsic point of view on hyperbolic planes. This is especially true because, as we will see, there is no standard embedding of a complete hyperbolic plane into 3-space.

A SHORT HISTORY OF HYPERBOLIC GEOMETRY

Hyperbolic geometry initially grew out of the Building Structures Strand through the work of János Bolyai (1802–1860, Hungarian), and N. I. Lobachevsky (1792–1856, Russian). Hyperbolic geometry is special from a formal axiomatic point of view because it satisfies all the postulates (axioms) of Euclidean geometry except for the parallel postulate. In hyperbolic geometry straight lines can converge toward each other

59

without intersecting (violating Euclid's Fifth Postulate, see Appendix A), and there is more than one straight line through a point that does not intersect a given line (violating the usual high school parallel postulate, which states that through any point P not on a given line l there is one and only one line through P not intersecting l). See Figure 5.1.

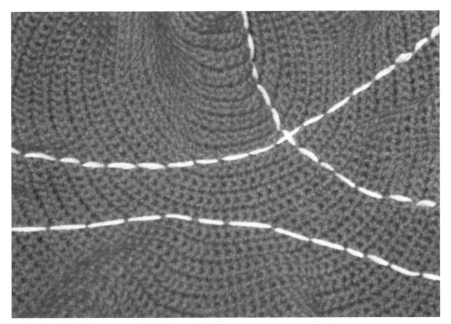

Figure 5.1   Two geodesics through a point not intersecting a given geodesic

The reader can explore more details of the axiomatic nature of hyperbolic geometry in Chapter 10. Note that the 450° cone also violates the two parallel postulates mentioned above. Thus the 450° cone has some of the properties of the hyperbolic plane.

Hyperbolic geometry has turned out to be useful in various branches of higher mathematics. Also, the geometry of binocular visual space appears experimentally to be best represented by hyperbolic geometry (see [**HY**: Zage]). In addition, hyperbolic geometry is one of the possible geometries for our three-dimensional physical universe — we will explore this connection more in Chapters 18 and 24.

Hyperbolic geometry and non-Euclidean geometry are considered in many books as being synonymous, but as we have seen there are other non-Euclidean geometries, especially spherical geometry. It is also not accurate to say (as many books do) that non-Euclidean geometry was discovered about 170 years ago. As we discussed in Chapter 2, spherical geometry (which is clearly not Euclidean) was in existence and studied (within the Navigation/Stargazing) by at least the ancient Babylonians, Indians, and Greeks more than 2000 years ago.

Most texts and popular books introduce hyperbolic geometry either axiomatically or via "models" of the hyperbolic geometry in the Euclidean plane. These models are like our familiar map projections of the surface of the earth. Like these maps of the earth's surface, intrinsic straight lines on the hyperbolic plane are not, in general, straight in the model (map) and the model, in general, distorts distances and angles. We will return to the subject of projection and models in Chapter 17. These "models" grew out of the Art/Pattern Strand.

In this chapter we will introduce the geometry of the hyperbolic plane as the intrinsic geometry of a particular surface in 3-space, in much the same way that we introduced spherical geometry by looking at the intrinsic geometry of the sphere in 3-space. This is more in the flavor of the Navigation/Stargazing Strand. Such a surface is called an *isometric embedding* of the hyperbolic plane into 3-space. We will construct such a surface in the next section. Nevertheless, many texts and popular books say that David Hilbert (1862–1943, German) proved in 1901 that it is not possible to have an isometric embedding of the hyperbolic plane onto a closed subset of Euclidean 3-space. These authors miss what Hilbert actually proved. In fact, Hilbert [**HY**: Hilbert] proved that there is no *real analytic* isometry (that is, no isometry defined by real-valued functions that have convergent power series). In 1964, N. V. Efimov [**HY**: Efimov] extended Hilbert's result by proving that there is no isometric embedding defined by functions whose first and second derivatives are continuous. Without giving an explicit construction, N. Kuiper [**HY**: Kuiper] showed in 1955 that there is a differentiable isometric embedding onto a closed subset of 3-space.

The construction used here was shown to David by William Thurston (b.1946, American) in 1978; and it is not defined by equations at all, because it has no definite embedding in Euclidean space. The idea for this construction is also included in [**DG**: Thurston], pages 49 and 50, and is discussed in [**DG**: Henderson], page 31. In Problem **5.3** we will show that our isometric model is locally isometric to a certain smooth surface of revolution called the *pseudosphere*, which is well known to locally have hyperbolic geometry. Later, in Chapter 17, we will explore the various (non-isometric) models of the hyperbolic plane (these models are the way that hyperbolic geometry is presented in most texts) and prove that these models and the isometric constructions here produce the same geometry.

DESCRIPTION OF ANNULAR HYPERBOLIC PLANES

In Appendix B we describe the details for five different isometric constructions of hyperbolic planes (or approximations to hyperbolic planes) as surfaces in 3-space. It is very important that you actually perform at least one of these constructions. The act of constructing the surface will give you a feel for hyperbolic planes that is difficult to get any other way. We will focus our discussions in the text on the description of the hyperbolic plane from annuli that was proposed by Thurston.

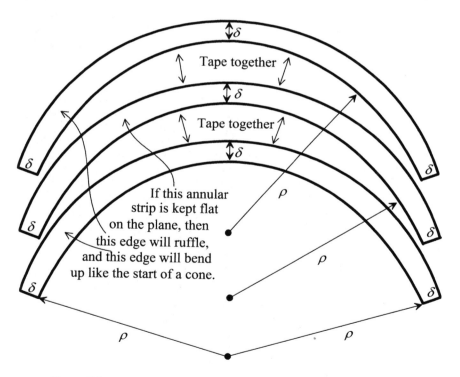

Figure 5.2   Annular strips for making an annular hyperbolic plane

A paper model of the hyperbolic plane may be constructed as follows: Cut out many identical annular ("annulus" is the region between two concentric circles) strips as in Figure 5.2. Attach the strips together by taping the inner circle of one to the outer circle of the other. It is crucial that all the annular strips have the same inner radius and the same outer radius, but the lengths of the annular strips do not matter. You can also cut an annular strip shorter or extend an annular strip by taping two strips together along their straight ends. The resulting surface is of course only an approximation of the desired surface. The actual hyperbolic plane is obtained by letting $\delta \to 0$ while holding the radius ρ fixed. Note that since the surface is constructed (as $\delta \to 0$) the same everywhere it is **homogeneous** (that is, intrinsically and geometrically, every point has a neighborhood that is isometric to a neighborhood of any other point). We will call the results of this construction the ***annular hyperbolic plane.*** I strongly suggest that the reader take the time to **cut out carefully several such annuli and tape them together as indicated.**

Daina discovered a process for crocheting the annular hyperbolic plane as described in Appendix B. The result is pictured in Figures 5.1 and 5.3 and other photos in this chapter.

Figure 5.3   A crocheted annular hyperbolic plane

There is also a polyhedral construction of the hyperbolic plane that is not directly related to the annular constructions but is easier for students (and teachers!) to construct. This construction (invented by David's son Keith Henderson) is called the hyperbolic soccer ball. See Appendix B for the details of the constructions (and templates) and Figure 5.4 for a picture. It also has a nice appearance if you make the heptagons a different color from the hexagons. As with any polyhedral construction we cannot get closer and closer approximations to the hyperbolic plane. There is also no apparent way to see the annuli.

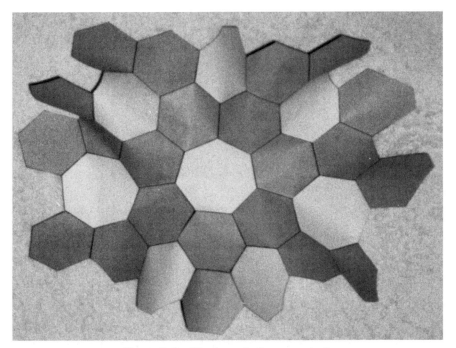

Figure 5.4   The hyperbolic soccer ball

HYPERBOLIC PLANES OF DIFFERENT RADII (CURVATURE)

Note that the construction of a hyperbolic plane is dependent on ρ (the radius of the annuli), which we will call the ***radius of the hyperbolic plane***. As in the case of spheres, we get different hyperbolic planes depending on the value of ρ. In Figures 5.5–5.7 there are crocheted hyperbolic planes with radii approximately 4 cm, 8 cm, and 16 cm. The pictures were all taken from approximately the same perspective and in each picture there is a centimeter rule to indicate the scale.

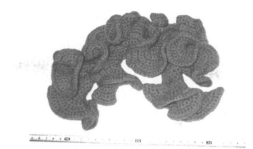

Figure 5.5   Hyperbolic plane with $\rho \approx 4$ cm

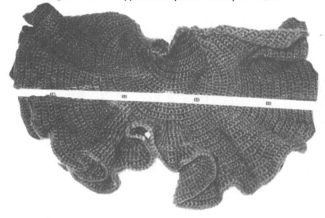

Figure 5.6   Hyperbolic plane with $\rho \approx 8$ cm

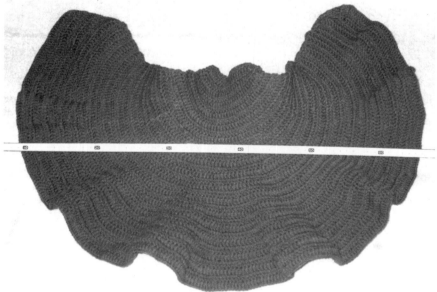

Figure 5.7   Hyperbolic plane with $\rho \approx 16$ cm

65

Note that as ρ increases, a hyperbolic plane becomes flatter and flatter (has less and less curvature). Both spheres and hyperbolic planes, as ρ goes to infinity, become indistinguishable from the ordinary flat (Euclidean) plane. Thus, the plane can be called a sphere (or hyperbolic plane) with infinite radius. In Chapter 7, we will define the *Gaussian Curvature* and show that it is equal to $1/\rho^2$ for a sphere and $-1/\rho^2$ for a hyperbolic plane.

PROBLEM 5.1   WHAT IS STRAIGHT IN A HYPERBOLIC PLANE?

a. *On a hyperbolic plane, consider the curves that run radially across each annular strip. Argue that these curves are intrinsically straight. Also, show that any two of them are asymptotic, in the sense that they converge toward each other but do not intersect.*

Look for the local intrinsic symmetries of each annular strip and then global symmetries in the whole hyperbolic plane. Make sure you give a convincing argument why the symmetry holds in the limit as $\delta \to 0$.

We shall say that two geodesics that converge in this way are ***asymptotic geodesics***. Note that there are no geodesics (straight lines) on the plane that are asymptotic.

b. *Find other geodesics on your physical hyperbolic surface. Use the properties of straightness (such as symmetries) you talked about in Problems 1.1, 2.1, and 4.1.*

Try holding two points between the index fingers and thumbs on your two hands. Now pull gently — a geodesic segment with its reflection symmetry should appear between the two points. If your surface is durable enough, try folding the surface along a geodesic. Also, you may use a ribbon to test for geodesics.

c. *What properties do you notice for geodesics on a hyperbolic plane? How are they the same as geodesics on the plane or spheres, and how are they different from geodesics on the plane and spheres?*

Explore properties of geodesics involving intersecting, uniqueness, and symmetries. Convince yourself as much as possible using your model — full proofs for some of the properties will have to wait until Chapter 17.

*Problem 5.2   Coordinate System on Annular Hyperbolic Plane

First, we will define coordinates on the annular hyperbolic plane that will help us to study it in Chapter 17. Let ρ be the fixed inner radius of the annuli and let H_δ be the approximation of the annular hyperbolic plane constructed from annuli of radius ρ and thickness δ. On H_δ pick the inner curve of any annulus, calling it the *base curve*; and on this curve pick any point as the origin O and pick a positive direction on this curve. We can now construct an (intrinsic) coordinate system $x_\delta \colon R^2 \to H_\delta$ by defining $x_\delta(0,0) = O$, $x_\delta(w,0)$ to be the point on the base curve at a distance w from O, and $x_\delta(w,0)$ to be the point at a distance s from $x_\delta(w,0)$ along the radial geodesic through $x_\delta(w,0)$, where the positive direction is chosen to be in the direction from outer to inner curve of each annulus. Such coordinates are often called ***geodesic rectangular coordinates***. See Figure 5.8.

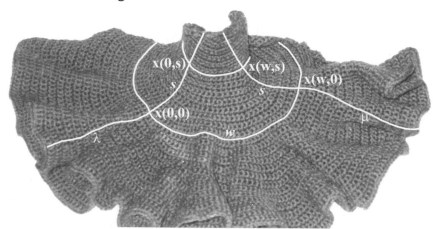

Figure 5.8   Geodesic rectangular coordinates on annular hyperbolic plane

a. *Show that the coordinate map x is one-to-one and onto from the whole of R^2 onto the whole of the annular hyperbolic plane. What maps to the annular strips, and what maps to the radial geodesics?*

b. *Let λ and μ be two of the radial geodesics described in part **a**. If the distance between λ and μ along the base curve is w, then show that the distance between them at a distance $s = n\delta$ from the base curve is, on the paper hyperbolic model,*

$$w\left(\frac{\rho}{\rho+\delta}\right)^{n} = w\left(\frac{\rho}{\rho+\delta}\right)^{s/\delta}.$$

Now take the limit as $\delta \to 0$ to show that the distance between λ and μ on the annular hyperbolic plane is

$$w \exp(-s/\rho).$$

Thus, the coordinate chart *x* preserves (does not distort) distances along the (vertical) second coordinate curves but at $x(a, b)$ the distances along the first coordinate curve are distorted by the factor of $\exp(-b/\rho)$ when compared to the distances in R^2.

*PROBLEM 5.3   THE PSEUDOSPHERE IS HYPERBOLIC

*Show that locally the annular hyperbolic plane is isometric to portions of a (smooth) surface defined by revolving the graph of a continuously differentiable function of z about the z-axis. This is the surface usually called the **pseudosphere**.*

OUTLINE OF PROOF

1. Argue that each point on the annular hyperbolic plane is like any other point. (Think of the annular construction. About a point consider a neighborhood that keeps *its* size as the width of the *annular strips*, δ, shrinks to zero.)

2. Start with one of the annular strips and complete it to a full annulus in a plane. Then construct a surface of revolution by attaching to the inside edge of this annulus other annular strips as described in the construction of the annular hyperbolic plane. (See Figure 5.9.) Note that the second and subsequent annuli form truncated cones. Finally, imagine the width of the annular strips, δ, shrinking to zero.

3. Derive a differential equation representing the coordinates of a point on the surface using the geometry inherent in Figure 5.9. If $f(r)$ is the height (*z*-coordinate) of the surface at a distance of *r* from

the z-axis, then the differential equation should be (remember that ρ is a constant)

$$\frac{dr}{dz} = \frac{-r}{\sqrt{\rho^2 - r^2}}.$$

Why?

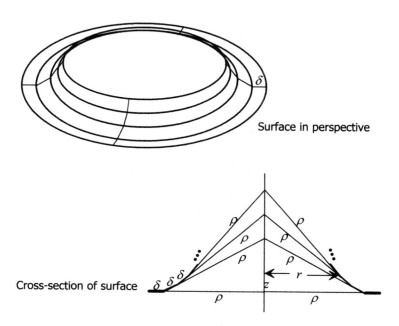

Surface in perspective

Cross-section of surface

Figure 5.9   Hyperbolic surface of revolution — pseudosphere

4. Solve (using tables or computer algebra systems) the differential equation for $z = f(r)$ as a function of r. Note that you are not getting r as a function of z. This curve is usually called the ***tractrix***.

5. Then argue (using a theorem from first-semester calculus) that r is a continuously differentiable function of z.

We can also crochet a pseudosphere by starting with 5 or 6 chain stitches and continuing in a spiral fashion, increasing as when crocheting the hyperbolic plane. See Figure 5.10. Note that, when you crochet beyond the annular strip that lays flat and forms a complete annulus, the surface forms ruffles and is no longer a surface of revolution (nor smooth).

Figure 5.10   Crocheted pseudosphere with ruffles

The term "pseudosphere" seems to have originated with Hermann von Helmholtz (1821–1894, German), who was contrasting spherical space with what he called pseudospherical space. However, Helmholtz did not actually find a surface with this geometry. Eugenio Beltrami (1835–1900, Italian) actually constructed the surface that is called the pseudosphere and showed that its geometry is locally the same as (locally isometric to) the hyperbolic geometry constructed by Lobatchevsky. (For more historical discussion, see [**HI**: Katz], pages 781–783.) Mathematicians searched further for a surface (in those days "surface" meant "real analytic surface") that would be the whole of the hyperbolic plane (as opposed to only being locally isometric to it). This search was halted when Hilbert proved that such a surface was impossible (in his theorem that we discussed above at the end of the first section in this chapter, A Short History of Hyperbolic Geometry).

Intrinsic/Extrinsic, Local/Global

On the plane or on spheres, rotations and reflections are both ***intrinsic*** in the sense that they are experienced by a 2-dimensional bug as rotations and reflections. These intrinsic rotations and reflections are also ***extrinsic*** in the sense that they can also be viewed as isometries of 3-space. (For example, the reflection of a sphere through a great circle can also be viewed as a reflection of 3-space through the plane containing the great circle.) Thus rotations and reflections are particularly easy to see on planes and spheres. In addition, on the plane and sphere all rotations and reflections are ***global*** in the sense that they take the whole plane to itself or whole sphere to itself. (For example, any intrinsic rotation about a point on a sphere is always a rotation of the whole sphere.) On cylinders and cones, intrinsic rotations and reflections exist locally because cones and cylinders are locally isometric with the plane. However, some intrinsic rotations on cones and cylinders are extrinsic and global: for example, rotations about the cone point on a circular cone with cone angle <360°, or half turns about any point on a cylinder. Rotations about the cone point on (>360°)-cones are global but not extrinsic. Rigid-motion-along-geodesic symmetries are extrinsic and global on cylinders but are neither on any cone. (Do you see why?) Reflections, in general, are neither extrinsic nor global (Can you see the exceptions on cones and cylinders?).

Problem 5.4   Rotations and Reflections on Surfaces

We can see from our physical hyperbolic planes that geodesics exist joining every pair of points and that these geodesics each have reflection-in-themselves symmetry. (If you did not see this in Problem **5.1c**, then go back and explore some more with your physical model. In Chapter 17 we will prove rigorously that this is in fact true by using the upper half-plane model.) In Chapter 17 we will show that these reflections are global reflections of the whole hyperbolic space. Note that there do not exist extrinsic reflections of the hyperbolic plane (embedded in Euclidean 3-space). Moreover, given all this, it is not clear that there exist intrinsic rotations, nor is it necessarily clear what exactly intrinsic rotations are.

a. *Let l and m be two geodesics on the hyperbolic plane that inter-sect at the point P. Look at the composition of the reflection R_l through l with the reflection R_m through m. Show that this composition R_mR_l deserves to be called a rotation about P. What is the angle of the rotation?*

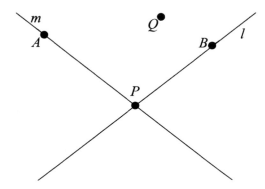

Figure 5.11   Composition of two reflections is a rotation

Let A be a point on m and B be a point on l, and let Q be an arbitrary point (not on m or l). Investigate where A, B, and Q are sent by R_l and then by R_mR_l. See Figure 5.11. Why are all points (except P) rotated through the same angle *and in the same direction.*

We will study symmetries and isometries in more detail in Chapter 11. In that chapter we will show that every isometry (on the plane, spheres, and hyperbolic planes) is a composition of one, two, or three reflections.

*b. *Show that Problem **3.2** (VAT) holds on cylinders, cones (includ-ing the cone points), and hyperbolic planes.*

If you check your proof(s) of **3.2** and modify them (if necessary) to involve only symmetries, then you will be able to see that they hold also on the other surfaces.

c. *Define "rotation of a figure about P through an angle θ" without mentioning reflections in your definition.*

What does a rotation do to a point not at P?

d. *A popular high school text **defines** a rotation as the composition of two reflections. Is this a good definition? Why or why not?*

Chapter 6

TRIANGLES AND CONGRUENCIES

> ***Polygons*** are those figures whose boundaries are made of straight lines: ***triangles*** being those contained by three, ...
>
> Things which coincide with one another are equal to one another.
>
> — Euclid, *Elements*, Definition 19 & Common Notion 4 [Appendix A]

At this point, you should be thinking intrinsically about the surfaces of spheres, cylinders, cones, and hyperbolic planes. In the problems to come you will have opportunities to apply your intrinsic thinking when you make your own definitions for triangle on these surfaces and investigate congruence properties of triangles.

In this chapter we will begin our study of triangles and their congruencies on all the surfaces that you have studied: plane, spheres, cones, cylinders, and hyperbolic spaces. (If you skipped any of these surfaces, you should find that this and the succeeding chapters will still make sense, but you will want to limit your investigations to triangles on the surfaces you studied.)

Before starting with triangles we must first discuss a little more general information about geodesics.

*GEODESICS ARE LOCALLY UNIQUE

In previous chapters we have studied geodesics, intrinsically straight paths. Our main criterion has been (in Chapter 2, 4, and 5) that a path is intrinsically straight (and thus, a geodesic) if it has local intrinsic reflection-through-itself symmetry. Using this notion, we found that joining any pair of points there is a geodesic that, on a sphere, is a great circle and, on a hyperbolic space, has reflection-through-itself symmetry. However, on more general surfaces, which may have no (even local) reflections, it is necessary to have a deeper definition of geodesic in terms of intrinsic curvature. (See for example, Chapter 3 of [**DG**: Henderson].) Then, to be precise, we must prove that the geodesics we found on spheres and hyperbolic planes are the only geodesics on these surfaces. It is easy to see that these geodesics that we have found are enough to give, for every point and every direction from that point, one geodesic proceeding from that point in that direction. To prove that these are the *only* geodesics, it is necessary (as we have mentioned before) to involve some notions from differential geometry. In particular, we must first define a notion of geodesic that will work on general surfaces that have no (even local) intrinsic reflections. Then we show that a geodesic satisfies a second-order (nonlinear) differential equation (see Problem **8.4b** of [**DG**: Henderson]). Thus, it follows from the analysis theorem on the *existence and uniqueness of differential equations*, with the initial conditions being a point on the geodesic and the direction of the geodesic at that point, that

> **THEOREM 6.0.** *For any given point and any direction at that point on a smooth surface there is a unique geodesic starting at that point and going in the given direction.*

From this it follows that the geodesics with local intrinsic reflection-in-itself symmetry, which we found in Problems **2.1**, **4.1**, and **5.1**, are all the geodesics on spheres, cylinders, cones, and hyperbolic planes.

PROBLEM 6.1   PROPERTIES OF GEODESICS

In this problem we ask you to pull together a summary of the properties of geodesics on the plane, spheres, and hyperbolic planes. Mostly, you have already argued that these are true, but we summarize the results here to remind us what we have seen and so that you can reflect again about why these are true. Remember that cylinders and cones (not at the cone point) are locally the same geometrically as (locally isometric to) the plane; thus, geodesics on the cone and cylinder are locally (but not globally) the same as straight lines on the plane.

> **a.** *For every geodesic on the plane, sphere, and hyperbolic plane there is a reflection of the whole space through the geodesic.*

b.   *Every geodesic on the plane, sphere, and hyperbolic plane can be extended indefinitely* (in the sense that the bug can walk straight ahead indefinitely along any geodesic).

c.   *For every pair of distinct points on the plane, sphere, and hyperbolic plane there is a* (not necessarily unique) *geodesic containing them.*

d.   *Every pair of distinct points on the plane or hyperbolic plane determines a unique geodesic segment joining them. On the sphere there are always at least two such segments.*

e.   *On the plane or on a hyperbolic plane, two geodesics either coincide or are disjoint or they intersect in one point. On a sphere, two geodesics either coincide or intersect exactly twice.*

Note that for the plane and hyperbolic plane, parts **d** and **e** are equivalent in the sense that they each imply the other.

Notice that these properties distinguish a sphere from both the Euclidean plane and from a hyperbolic plane; however, these properties do not distinguish the plane from a hyperbolic plane.

PROBLEM **6.2**   ISOSCELES TRIANGLE THEOREM (ITT)

In order to start out with some common ground, let us agree on some terminology: A *triangle* is a geometric figure formed of three points (*vertices*) joined by three straight line (geodesic) segments (*sides*). A triangle divides the surface into two regions (the *interior* and *exterior*). The (*interior*) *angles* of the triangle are the angles between the sides in the interior of the triangle. (As we will discuss below, on a sphere you must decide which region you are going to call the interior — often the choice is arbitrary.)

We will find the Isosceles Triangle Theorem very useful in studying circles and the other congruence properties of triangles because the two congruent sides can be considered to be radii of a circle.

a.   (**ITT**) *Given a triangle with two of its sides congruent, then are the two angles opposite those sides also congruent? See* Figure 6.1. *Look at this on all five of the surfaces we are studying.*

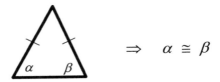

Figure 6.1   ITT

Use symmetries to solve this problem. First, look at this on the plane and note what properties of the plane you use. Then look on other surfaces. Look for counterexamples — if there were counterexamples, what could they look like? If you think that ITT is not true for all triangles on a particular surface, then describe a counterexample **and** look for a smaller class of triangles that do satisfy ITT on that surface. In the process of these investigations you will need to use properties of geodesics on the various surfaces (see Problem **6.1**). State explicitly what properties you are using. Hint: On a sphere two given points do not determine a unique geodesic segment but two given points plus a third point collinear to the given two **do** determine a unique geodesic segment.

In your proof of Part **a**, try to see that you have also proved the following very useful result:

b. COROLLARY. *The bisector of the top angle of an isosceles triangle is also the perpendicular bisector of the base of that triangle.*

You may also want to prove a converse of ITT, but we will use it in this book only in Problem **14.4**:

c. CONVERSE OF **ITT**. *If two angles of a triangle are congruent, then are the sides opposite these angles also congruent?*

Use symmetry and look out for counterexamples — they do exist for the converse.

CIRCLES

To study congruencies of triangles we will need to know something about circles and constructions of bisectors and perpendicular bisectors.

We define a **circle** intrinsically: *A circle with center P and radius PQ is the collection of all points X which are connected to P by a segment PX which is congruent to PQ.*

Note that on a sphere every circle has two (intrinsic) centers that are antipodal (and, in general, two different radii). See Figure 6.2.

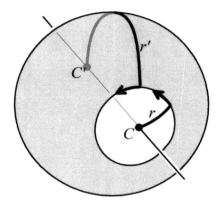

Figure 6.2   Circles on a sphere have two centers

Now ITT can be used to prove theorems about circles. For example,

THEOREM 6.2. *On the plane, spheres, hyperbolic planes, and locally on cylinders and cones, if the centers of two circles are disjoint (and not antipodal), then the circles intersect in either 0, 1, or 2 points. If the centers of the two circles coincide (or are antipodal), then the circles either coincide or are disjoint.*

Proof: Because cylinders and cones are locally isometric to the plane, locally and intrinsically circles will behave the same as on the plane. Thus we limit the remainder of this proof to the plane, spheres, and hyperbolic planes. Let C and C' denote the centers of the circles. See Figure 6.3.

If C and C' are antipodal on a sphere and the two circles intersect at P, then a (extrinsic) plane through P (perpendicular to the extrinsic diameter CC') will intersect the sphere in a circle that must coincide with the two given circles. If C and C' coincide on a sphere and the circles intersect, then pick the antipodal point to C as the center of the first circle, which reduces this to the case we just considered. If C and C' coincide on the plane and hyperbolic planes and the circles intersect, then the circles have the same radii because two points are joined by only one line segment. Thus, if the centers coincide or are antipodal, the circles coincide or are disjoint.

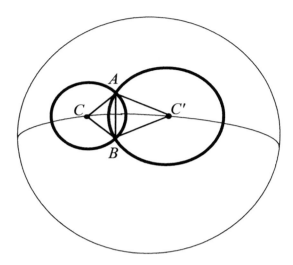

Figure 6.3   Intersection of two circles

Thus, we can now assume that C and C' are disjoint and not antipodal, so there is a unique geodesic joining the centers. If A and B are two points of intersection of the circles, then $\triangle ACB$ and $\triangle AC'B$ are isosceles triangles.

But given that $\triangle ACB$ and $\triangle AC'B$ are isosceles, the Corollary to ITT asserts that the bisectors of $\angle ACB$ and $\angle AC'B$ must be perpendicular bisectors of their common base. Thus, the union of the two angle bisectors is straight and joins C and C'. So the union must be contained in the unique geodesic determined by C and C'. Therefore, any pair of intersections of the two circles, such as A and B, must lie on opposite sides of this unique geodesic. Immediately, it follows that there cannot be more than two intersections.

TRIANGLE INEQUALITY

On the plane, we have the following well-known result:

> ***Triangle Inequality on Plane or a Hyperbolic Plane:*** *The combined length of any two sides of a triangle is greater than the length of the third side.*

Do you see how this follows from our discussion of circles? See Figure 6.3.

The Triangle Inequality is a partial expression of the statement that "a straight line is the shortest distance between two points" and because of that we might expect that the triangle inequality is false on spheres (and cylinders and cones). *Can you find a counterexample?* But we can make a simple change that makes it work on the sphere:

> ***Triangle Inequality on a Sphere:*** *The combined lengths of any two sides is not less than the (shortest) distance between the end points of the third side.*

This leads to a definition that will be useful in some later chapters:

> **Definition.** *For any line segment, l, with endpoints, A, B, we define the (**special**) **absolute value** of l (in symbols, $|l|_s$) to be the shortest distance from A to B.*

Note that on the plane (or in a vector space) this is the same as the usual "absolute value" (or "norm").

PROBLEM 6.3   BISECTOR CONSTRUCTIONS

    a. *Show how to use a compass and straightedge to construct the perpendicular bisector of a straight line segment. How do you know it is actually the perpendicular bisector? How does it work on the sphere and hyperbolic plane?*

Use ITT and Theorem **6.2**. Be sure that you have considered all segment lengths on the sphere. Hint: Use Figures 6.3 and the arguments in the section Circles.

    b. *Show how to use a compass and straightedge to find the bisector of any planar angle. How do you know it actually is the angle bisector? How does it work on the sphere and hyperbolic plane?*

Use ITT and part **a**. Be sure that you have considered all sizes of angles.

It is a part of mathematical folklore that it is impossible to trisect an angle with compass and straightedge; however, you will show in Problem **15.4** that, in fact, it is possible. In addition, we will discuss what is a correct statement of the impossibility of trisecting angles.

PROBLEM 6.4   SIDE-ANGLE-SIDE (SAS)

We now investigate properties that will allow us to say that two triangles are "the same". Let us clarify some terminology that we have found to be helpful for discussing SAS and other theorems. Two triangles are said to be *congruent* if, through a combination of translations, rotations, and reflections, one of them can be made to coincide with the other. In fact (as we will prove in Chapter 11), we only need to use reflections. If an even number of reflections are needed, then the triangles are said to be *directly congruent*, because in this case (as we show in Problem **11.3**) the reflections can be replaced pairwise by rotations and translations. In this text we will focus on *congruence* and not specifically on *direct congruence*; however, some readers may wish to keep track of the distinction as we go along.

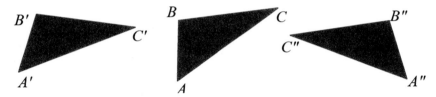

Figure 6.4   Direct congruence and congruence

In Figure 6.4, $\triangle ABC$ is directly congruent to $\triangle A'B'C'$ but $\triangle ABC$ is not directly congruent to $\triangle A''B''C''$. However, $\triangle ABC$ is congruent to both $\triangle A'B'C'$ and $\triangle A''B''C''$ and we write: $\triangle ABC \cong \triangle A'B'C' \cong \triangle A''B''C''$.

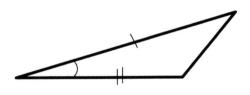

Figure 6.5   SAS

Are two triangles congruent if two sides and the included angle of one are congruent to two sides and the included angle of the other? See Figure 6.5.

In some textbooks SAS is listed as an axiom; in others it is listed as the definition of congruency of triangles, and in others as a theorem to be proved. But no matter how one considers SAS, it still makes sense and is important to ask, Why is SAS true on the plane?

Is SAS true on spheres, cylinders, cones, and hyperbolic planes?

If you find that SAS is not true for all triangles on a sphere or another surface, is it true for sufficiently small triangles? Come up with a definition for "small triangles" for which SAS does hold.

SUGGESTIONS

Be as precise as possible, but **use your intuition**. In trying to prove SAS on a sphere you will realize that SAS does not hold unless some restrictions are made on the triangles. Keep in mind that everyone sees things differently, so there are many possible definitions of "small." Some may be more restrictive than others (that is, they don't allow as many triangles as other definitions). Use whatever definition makes sense for you.

Remember that it is not enough to simply state what a small triangle is; you must also prove that SAS is true for the small triangles under your definition — explain why the counterexamples you found before are now ruled out and explain why the condition(s) you list is (are) sufficient to prove SAS. Also, try to come up with a basic, general proof that can be applied to all surfaces.

And remember what we said before: By "proof" we mean what most mathematicians use in their everyday practice, that is, *a convincing communication that answers — Why?* We do not ask for the two-column proofs that used to be common in North American high schools (unless, of course, you find the two-column proof is sufficiently convincing and answers — Why?). Your proof should convey the meaning you are experiencing in the situation. Think about why SAS is true on the plane — think about what it means for actual physical triangles — then try to translate these ideas to the other surfaces.

So why is SAS true on the plane? We will now illustrate one way of looking at this question. Referring to Figure 6.6, suppose that $\triangle ABC$ and $\triangle A'B'C'$ are two triangles such that $\angle BAC \cong \angle B'A'C'$, $AB \cong A'B'$ and $AC \cong A'C'$. Reflect $\triangle A'B'C'$ about the perpendicular bisector (Problem **6.3**) of AA' so that A' coincides with A. Because the sides AC and $A'C'$ are congruent, we can now reflect about the angle bisector of $\angle C'AC$. Now C' coincides with C. (*Why?*) If after this reflection B and B' are not coincident, then a reflection (about $AC = A'C'$) will complete the process and all three vertices, the two given sides, and the included angle of the two triangles will coincide. *So why is it that, on the plane, the third sides (BC and B'C') must now be the same?*

Step 1: Two triangles with SAS

Step 2: Reflect about the
perpendicular bisector of AA'

Step 3: Reflect about the
angle bisector of angle $C'AC$

Step 4: Reflect about AC

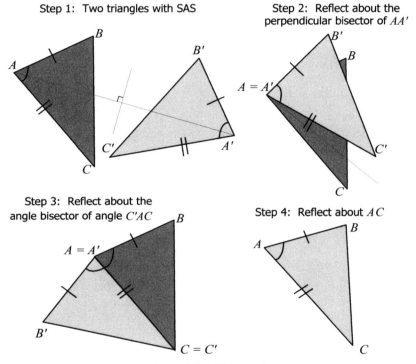

Figure 6.6   SAS on plane

Because the third sides (*BC* and *B'C'*) coincide, $\triangle ABC$ is congruent to $\triangle A'B'C'$. (In the case that only two reflections are needed, the two triangles are directly congruent.)

The proof of SAS on the plane is not directly applicable to the other surfaces because properties of geodesics differ on the various surfaces. In particular, the number of geodesics joining two points varies from surface to surface and is also relative to the location of the points on the surface. On a sphere, for example, there are always at least two straight paths joining any two points. As we saw in Chapter 4, the number of geodesics joining two points on a cylinder is infinite. On a cone the number of geodesics is dependent on the cone angle, but for cones with angles less than 180° there is more than one geodesic joining two points. It follows that the argument made for SAS on the plane is not valid on cylinders, cones, or spheres. The question then arises: Is SAS *ever* true on those surfaces?

Look for triangles for which SAS is not true. Some of the properties that you found for geodesics on spheres, cones, cylinders, and hyperbolic planes will come into play. As you look closely at the features of triangles on those surfaces, you may find that they challenge your notions of triangle. Your intuitive notion of triangle may go beyond what can be put into a traditional definition of triangle. When you look for a definition of *small triangle* for which SAS will hold on these surfaces, you should try to stay close to your intuitive notion. In the process of exploring different triangles you may come up with examples of triangles that seem very strange. Let us look at some unusual triangles.

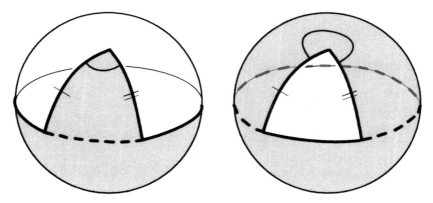

Figure 6.7   Two counterexamples for SAS on sphere

For instance, keep in mind the examples in Figure 6.7. All the lines shown in Figure 6.7 are geodesic segments of the sphere. The two sides and their included angle for SAS are marked. As you can see, there are two possible geodesics that can be drawn for the third side — the short one in front and the long one that goes around the back of the sphere. Remember that on a sphere, any two points define at least two geodesics (an infinite number if the points are at opposite poles).

Look for similar examples on a cone and cylinder. You may decide to accept the smaller triangle into your definition of "small triangle" but to exclude the large triangle from your definition. But what is a large triangle? To answer this, let us go back to the plane. What is a triangle on the plane? What do we choose as a triangle on the plane?

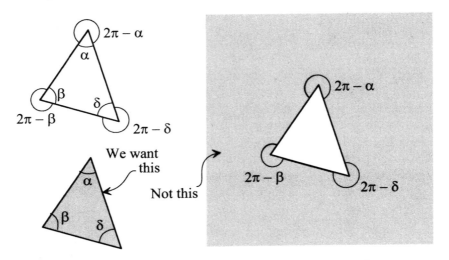

Figure 6.8   We choose the interior of a plane triangle to have finite area

On the plane, a figure that we want to call a triangle has all of its angles on the "inside." Also, there is a clear choice for *inside* on the plane; it is the side that has finite area. See Figure 6.8. But what is the inside of a triangle on a sphere?

The restriction that the area on the inside has to be finite does not work for the spherical triangles because all areas on a sphere are finite. So what is it about the large triangle that challenges our view of triangle? You might try to resolve the triangle definition problem by specifying that each side must be the shortest geodesic between the endpoints. However, be aware that antipodal points (that is, a pair of points that are at diametrically opposite poles) on a sphere do not have a unique shortest geodesic joining them. On a cylinder we can have a triangle for which all the sides are the shortest possible segments, yet the triangle does not have finite area. Try to find such an example. In addition, a triangle on a cone will always bound one region that has finite area, but a triangle that encircles the cone point may cause problems.

PROBLEM 6.5 ANGLE-SIDE-ANGLE (ASA)

Are two triangles congruent if one side and the adjacent angles of one are congruent to one side and the adjacent angles of another? See Figure 6.9.

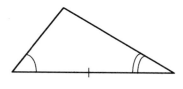

Figure 6.9   ASA

SUGGESTIONS

This problem is similar in many ways to the previous one. As before, look for counterexamples on all surfaces; and if ASA does not hold for all triangles, see if it works for small triangles. If you find that you must restrict yourself to small triangles, see if your previous definition of "small" still works; if it does not work here, then modify it.

It is also important to keep in mind when considering ASA that both of the angles must be on the same side — the *interior* of the triangle. For example, see Figure 6.10.

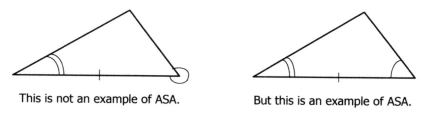

This is not an example of ASA.   But this is an example of ASA.

Figure 6.10   Angles of a triangle must be on same side

Let us look at a proof of ASA on the plane as depicted in Figure 6.11.

The planar proof for ASA does not work on spheres, cylinders, and cones because, in general, geodesics on these surfaces intersect in more than one point. But can you make the planar proof work on a hyperbolic plane?

Step 1: Two triangles with ASA

Step 2: Reflect about the
perpendicular bisector

Step 3: Reflect about the
angle bisector

Step 4: Reflect about the
congruent side

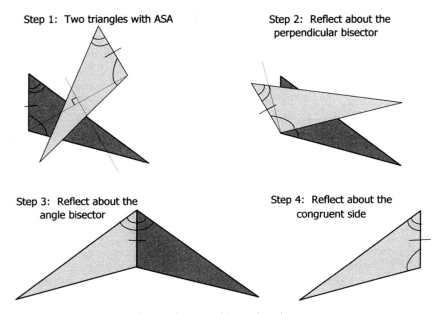

Figure 6.11   ASA on the plane

As was the case for SAS, we must ask ourselves if we can find a class of small triangles on each of the different surfaces for which the above argument is valid. You should check if your previous definitions of small triangle are too weak, too strong, or just right to make ASA true on spheres, cylinders, cones, and hyperbolic planes. It is also important to look at cases for which ASA does not hold. Just as with SAS, some interesting counterexamples arise.

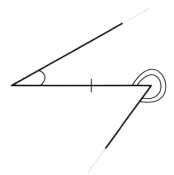

Figure 6.12   Possible counterexample to ASA

In particular, try out the configuration in Figure 6.12 on a sphere. To see what happens you will need to try this on an actual sphere. If you extend the two sides to great circles, what happens? You may instinctively say that it is not possible for this to be a triangle, and on the plane most people would agree, but *try it on an physical sphere and see what happens*. Does it define a unique triangle? Remember that on a sphere two geodesics always intersect twice.

Finally, notice that in our proof of ASA on the plane, we did not use the fact that the sum of the angles in a triangle is 180°. We avoided this for two reasons. For one thing, to use this "fact" we would have to prove it first. This is both time consuming and unnecessary. We will prove it later (in different ways) in Chapters 7 and 10. More importantly, such a proof will not work on spheres and hyperbolic planes because the sum of the angles of triangles on spheres and hyperbolic planes is not always 180° — see the triangles depicted in Figures 6.13 and 6.14. We will explore further the sum of the angles of a triangle in Chapter 7.

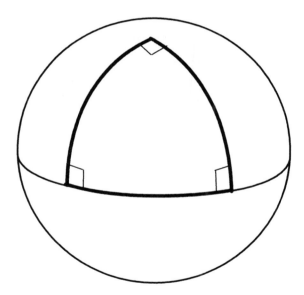

Figure 6.13   Triple-right triangle on a sphere

Remember that it is best to come up with a proof that will work for all surfaces because this will be more powerful, and, in general, will tell us more about the relationship between the plane and the other surfaces.

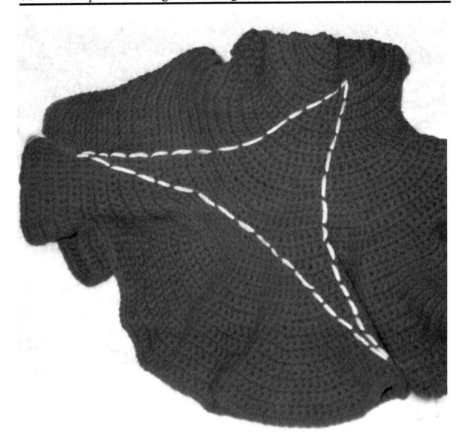

Figure 6.14   Hyperbolic triangle

Chapter 7

AREA AND HOLONOMY

We [my student and I] are both greatly amazed; and my share
in the satisfaction is a double one, for he sees twice over who
makes others see.

— Jean Henri Fabre, *The Life of the Fly*,
New York: Dodd, Mead and Co., 1915, p. 300.

There are many things in this chapter that have amazed us and our
students. We hope you, the reader, will also be amazed by them. We will
find a formula for the area of triangles on spheres and hyperbolic planes.
We will then investigate the connections between area and *parallel
transport*, a notion of local parallelism that is definable on all surfaces.
We will also introduce the notion of *holonomy*, which has many applications in modern differential geometry and engineering.

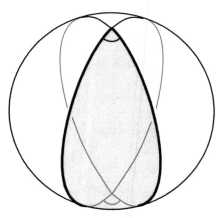

Figure 7.1   Lune or biangle

DEFINITION: A *lune* or *biangle* is any of the four regions determined by two (not coinciding) great circles (see Figure 7.1). The two angles of the lune are congruent. (*Why?*)

PROBLEM 7.1   THE AREA OF A TRIANGLE ON A SPHERE

a. *The two sides of each interior angle of a triangle Δ on a sphere determine two congruent lunes with lune angle the same as the interior angle. Show how the three pairs of lunes determined by the three interior angles, α, β, γ, cover the sphere with some overlap. (What is the overlap?)*

Draw this on a physical sphere, as in Figure 7.2.

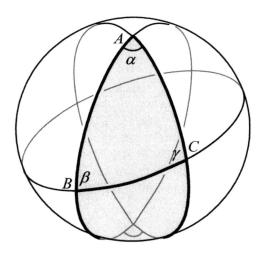

Figure 7.2   Finding the area of a spherical triangle

b. *Find a formula for the area of a lune with lune angle θ in terms of θ and the (surface) area of the sphere (of radius ρ), which you can call S_ρ. Use radian measure for angles.*

Hint: What if θ is π? π/2?

c. *Find a formula for the area of a triangle on a sphere of radius ρ.*

SUGGESTIONS

This is one of the problems that you almost certainly must do on an actual sphere. There are simply too many things to see, and the drawings we make on paper distort lines and angles too much. The best way to start is to make a small triangle on a sphere and extend the sides of the triangle to complete great circles. Then look at what you've got. You will find an identical triangle on the other side of the sphere, and you can see several lunes that extend out from the triangles. The key to this problem is to put everything in terms of areas that you know. We will see later (Problem **14.3**) that the area of the whole sphere with radius ρ is $S_\rho = 4\pi\rho^2$, or you may find a derivation of this formula in a multivariable calculus text, or you can just leave your answer in terms of S_ρ.

HISTORICAL NOTE

Formulas expressing the area of a spherical triangle and polygon in terms of their respective angular excesses appeared in print for the first time in the paper "On a newly discovered measure of area of spherical triangles and polygons" published as an appendix to *A new invention in algebra* (*Invention nouvelle en l'algebre*, Amsterdam, 1629) by the Flemish mathematician Albert Girard (1595–1632). For Girard's proof, see [**HI**: Rosenfeld], pp. 27–31. A proof similar to the one indicated here was first published in 1781 by the mathematician Leonhard Euler (1707–1783).

PROBLEM **7.2**   AREA OF HYPERBOLIC TRIANGLES

Before we start to explore the area of a general triangle on the hyperbolic plane, we first look for triangles with large area.

> **a.**   *On your hyperbolic plane draw as large a triangle as you can find. Compare your triangle with the large triangles that others have found. What do you notice?*

This part of the problem is best to do communicating with other people.

We can try to mimic the derivation of the area of spherical triangles, but of course there are no lunes and the area of the hyperbolic plane is evidently infinite. Nevertheless, if we focus on the exterior angles of a hyperbolic triangle and look at the regions formed, we obtain a picture of the situation in the annular hyperbolic plane. See Figure 7.3. Draw this picture on your hyperbolic plane.

In Figure 7.3, a triangle is drawn with its interior angles, α, β, γ, and exterior angles, $\pi-\alpha$, $\pi-\beta$, $\pi-\gamma$. The three extra lines are geodesics that are asymptotic at both ends to an extended side of the triangle. We call the region enclosed by these three extra geodesics an ***ideal triangle***. In the annular hyperbolic plane these are not actually triangles because their vertices are at infinity. In Figure 7.3 we see that the ideal triangle is divided into the original triangle and three "triangles" that have two of their vertices at infinity. We call a "triangle" with two vertices at infinity (and all sides geodesics) a **2/3-*ideal triangle***. You can use this decomposition to determine the area of a hyperbolic triangle in much the same way you determined the area of a spherical triangle. So first we must investigate the areas of ideal and 2/3-ideal triangles.

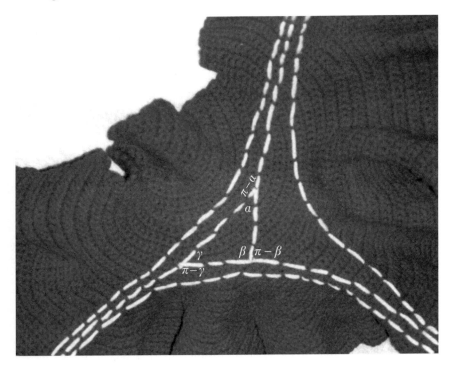

Figure 7.3   Triangle with an ideal triangle and three 2/3-ideal triangles

Now let us look at 2/3-ideal triangles.

b.   *Show that on the same hyperbolic plane, all 2/3-ideal triangles
with the same angle θ are congruent.*

Think of the proof of SAS (Problem **6.4**). If you have two 2/3-ideal triangles with angle θ, then by reflections you can place one of the θ-angles
on top of the other. The triangles will then coincide except possibly for
the third side, which is asymptotic to the two sides of the angle θ. Now
you must argue that these third sides must coincide. Or, in other words,
why is the situation in Figure 7.4 impossible for 2/3-ideal triangles on a
hyperbolic plane? Note, from Problem **5.4**, that we can rotate so that any
geodesic we pick is (after rotation) a radial geodesic. Problem **5.2** may
be helpful.

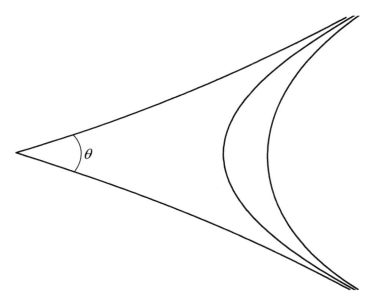

Figure 7.4   Are 2/3-ideal triangles with angle θ congruent?

Because all the 2/3-ideal triangles are congruent, we can define an
area function as

$$A_\rho(\alpha) = \text{area of a 2/3 ideal triangle with } \textit{exterior} \text{ angle } \alpha$$
$$\text{on a hyperbolic plane with radius } \rho.$$

c. *Show that the area function A_ρ is an additive function. That is,*

$$A_\rho(\alpha + \beta) = A_\rho(\alpha) + A_\rho(\beta).$$

Look at the picture in Figure 7.5 and show that the area of $\triangle ADE$ is the sum of the areas of triangles $\triangle ABC$ and $\triangle ACE$ by showing that $\triangle PDE$ is congruent to $\triangle PBC$.

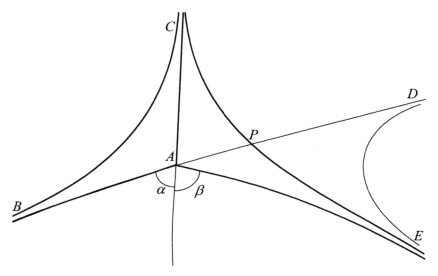

Figure 7.5   Area of 2/3-ideal triangles is additive

So we now have shown that the area function A_ρ is additive and it is also clearly continuous.

THEOREM 7.2. *A continuous additive function (from the real numbers to the real numbers) is linear.*

Because the area function is additive, it also is true that it is linear over the rational numbers. For example,

$$2A_\rho(\alpha) = A_\rho(\alpha) + A_\rho(\alpha) = A_\rho(\alpha + \alpha) = A_\rho(2\alpha),$$

and, if you set $\beta = 2\alpha$, then the same equations show that $\frac{1}{2}A_\rho(\beta) = A_\rho(\frac{1}{2}\beta)$. Thus, because the area function is continuous, the function must be linear (over the real numbers).

Therefore, $A_\rho(\alpha)$ = constant × α, for $0 \le \alpha < \pi$. We can conclude that $A_\rho(0) = 0$. If we let the finite vertex of 2/3-ideal triangle go to infinity, then the interior angle will go to zero and the exterior angle will go to π. Thus $A_\rho(\pi)$ must be the area of an ideal triangle. So we have proved the following:

All ideal triangles on the same hyperbolic plane have the same area, which we can call I_ρ.

So we can write the area function as

$$A_\rho(\alpha) = \alpha \times (I_\rho/\pi).$$

In fact, we will show in Problem **17.4** that all ideal triangles (on the same hyperbolic plane) are congruent. This is a result you may have guessed from your work in part **a**, you can also prove it using part **b**. We will also show in Problem **17.4** that the formula for the area of an ideal triangle is $I_\rho = \pi\rho^2$. Then it follows that $A_\rho(\alpha) = \alpha\rho^2$. Notice that it is only after **17.4** that we know for certain that 2/3-ideal triangles (and ideal triangles) have finite area, though you may have surmised that from your work on part **a**.

d.   *Find a formula for the area of a hyperbolic triangle.*

Look at Figure 7.3 and put it together with what we have just proved.

HISTORICAL NOTE
The proof is based on a proof that C. F. Gauss included in a 1832 letter to J. Bolyai's father that is published in his collected works.

PROBLEM 7.3   SUM OF THE ANGLES OF A TRIANGLE

a.   *What can you say about the sum of the interior angles of triangles on spheres and hyperbolic planes? Are there maximum and/or minimum values for the sum?*

Look at triangles with non-zero area and use your formulas from Problems **7.1** and **7.2**.

b.   *What is the sum of the (interior) angles of a planar triangle?*

Let △ABC be a triangle on the plane and imagine a sphere of radius ρ passing through the points A, B, C. These three points also determine a small spherical triangle on the sphere. Now imagine the radius ρ growing to infinity and the spherical triangle converging to the planar triangle.

This result for the plane is normally proved after invoking a parallel postulate. Here, we are making the assumption that the plane is a sphere of infinite radius. We will turn to a discussion of the various parallel postulates in Chapter 10.

INTRODUCING PARALLEL TRANSPORT

Imagine that you are walking along a straight line or geodesic carrying a horizontal stick that makes a fixed angle with the line you are walking on. If you walk along the line maintaining the direction of the stick relative to the line constant, then you are performing a ***parallel transport*** of that "direction" along the path. (See Figure 7.6.) "Parallel transport" is sometimes called "parallel displacement" or "parallel transfer".

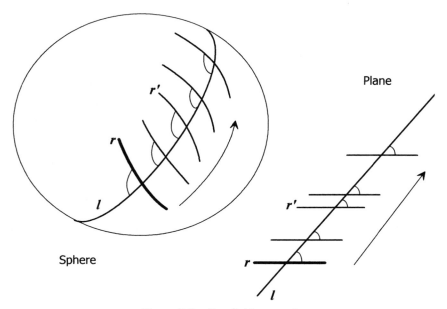

Figure 7.6   Parallel transport

To express the parallel transport idea, it is common terminology to say that

- *r'* is a parallel transport of *r* along *l*;

- *r* is a parallel transport of *r'* along *l*;

- *r* and *r'* are parallel transports along *l*;

- *r* can be parallel transported along *l* to *r'*; or

- *r'* can be parallel transported along *l* to *r*.

On the plane there is a global notion of parallelism — two lines in the same plane are ***parallel*** if they do not intersect when extended. As we will see in Problem **8.2** (or, for the plane, from standard results in high school geometry), if two lines are parallel transports along another line in the plane or the hyperbolic plane, then they are also parallel in the sense that they will not intersect if extended. On a sphere this is not true — any two great circles on the same sphere intersect and intersect twice. In Problem **10.1** you will show that if two lines in the plane are parallel transports along a third line, then they are parallel transports along every line that transverses them. This is also not true on a sphere and not true on a hyperbolic plane. For example, any two great circles (longitudes) through the north pole are parallel transports of each other along the equator, but they are not parallel transports along great circles near the north pole. We will explore this aspect of parallel transport more in Chapters 8 and 10.

Parallel transport has become an important notion in differential geometry, physics, and mechanics. One important aspect of differential geometry is the study of properties of spaces (surfaces) from an intrinsic point of view. As we have seen, it is not in general possible to have a global notion of direction that will determine when a direction (vector) at one point is the same as a direction (vector) at another point. However, we can say that they have the same direction *with respect to* a geodesic *g* if they are parallel transports of each other along *g*.

Parallel transport can be extended to arbitrary curves, as we shall discuss at the end of this chapter. There is even a mechanical device (first developed in third-century China!), called the "South-Seeking

Chariot," which will perform parallel transport along a curve on a surface. See [**DG**: Santander].

With this notion it is possible to talk about the rate at which a particular vector quantity changes intrinsically along a curve (covariant differentiation). In general, covariant differentiation is useful in the areas of physics and mechanics. In physics, the notion of parallel transport is central to some of the theories that have been put forward as possible candidates for a "unified field theory," a hoped-for but as yet unrealized theory that would unify all known physical laws about forces of nature.

HISTORICAL NOTE

According to [**HI**: Kline], p. 1132, parallel transport was first introduced in 1906 by L. E. J. Brouwer (1881–1966) in the context of surfaces that are locally Euclidean, spherical, or hyperbolic. The general notion of parallel transport was introduced in 1917 independently by Tullio Levi-Civita (1873–1941) and Gerhard Hessenberg (1874–1925).

INTRODUCING HOLONOMY

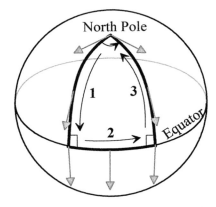

Figure 7.7   The holonomy of a double-right triangle on a sphere

Let us explore what happens when we parallel transport a line segment around a triangle. For example, consider on a sphere an isosceles triangle with base on the equator and opposite vertex on the north pole (see Figure 7.7). Note that the base angles are right angles. Now start at the north pole with a vector (a directed geodesic segment — the gray arrows in Figure 7.7) and parallel transport it along one of the sides of the triangle until it reaches the base. Then parallel transport it along the base to the third side. Then parallel transport back to the north pole along the third side. Notice that the vector now points in a different direction than it did originally. You can follow a similar story for the right hyperbolic triangle represented in Figure 7.8 and see that here also there is a difference between the starting vector and the ending parallel transported vector. This difference is called the ***holonomy*** of the triangle. Note that the difference angle is counterclockwise on the sphere and clockwise in the hyperbolic plane.

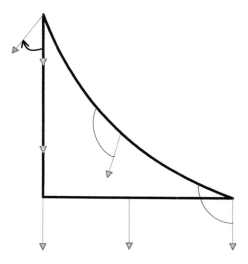

Figure 7.8   Holonomy of a hyperbolic triangle

This works for any small triangle (that is, a triangle that is contained in an open hemisphere) on a sphere and for all triangles in a hyperbolic plane. We can define ***the holonomy of a*** (***small***, if on a sphere) ***triangle***, $\mathcal{H}(\Delta)$, as follows:

If you parallel transport a vector (a directed geodesic segment) counterclockwise around the three sides of a small triangle, then the holonomy of the triangle is the smallest angle from the original position of the vector to its final position with counterclockwise being positive and clockwise being negative.

For the spherical triangle in Figure 7.7 we see that the holonomy is positive and equal to the upper angle of the triangle. For the hyperbolic triangle in Figure 7.8 we see that the holonomy is negative (clockwise).

Holonomy can also be defined for large triangles on a sphere, but it is more complicated because of the confusion as to what angle to measure. For example, what should be the holonomy when you parallel transport around the equator — 0 radians or 2π radians? Compare with the formula for the area of a spherical triangle from Problem **7.2**.

PROBLEM 7.4   THE HOLONOMY OF A SMALL TRIANGLE

Find a formula that expresses the holonomy of a small triangle on a sphere and a formula that expresses the holonomy of any triangle on a hyperbolic plane. What is the holonomy of a triangle on the plane?

SUGGESTIONS

What happens to the holonomy when you change the angle at the north pole of the triangle in Figure 7.7? What happens if you parallel transport around the triangle a vector pointing in a different direction? Parallel transport vectors around different triangles on your model of a sphere. Try it on triangles that are very nearly the whole hemisphere and try it on very small triangles. What do you notice? Try this also on your models of the hyperbolic plane, again for different size triangles.

A good way to approach the formula for general triangles is to start with any geodesic segment at one of the angles of the triangle and follow it as it is parallel transported around the triangle. Keep track of the relationships between the angles this segment makes with the sides and the exterior angles. See Figure 7.9, which is drawn for spherical triangles; the reader should be able to draw an analogous picture for a general hyperbolic triangle.

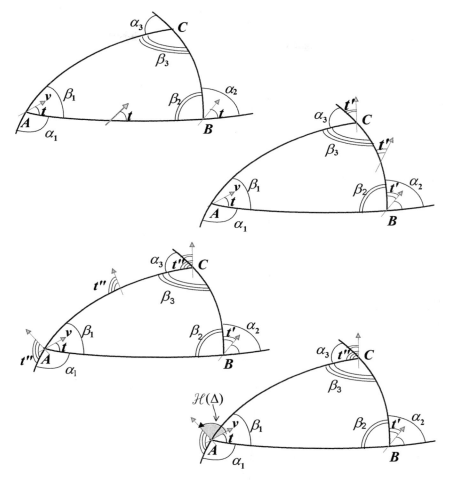

Figure 7.9   Holonomy of a general triangle

![pencil icon] **Pause, explore, and write out your ideas for this problem before reading further.**

THE GAUSS-BONNET FORMULA FOR TRIANGLES

In working on Problem **7.4** you should find (among other things) that

The holonomy of a (small, if on a sphere) triangle is equal to 2π minus the sum of the exterior angles or equal to the sum of the interior angles minus π.

Let β_1, β_2, β_3 be the interior angles of the triangle and α_1, α_2, α_3 the exterior angles. Then algebraically the statement above can be written as

$$\mathcal{H}(\Delta) = 2\pi - (\alpha_1 + \alpha_2 + \alpha_3) = (\beta_1 + \beta_2 + \beta_3) - \pi.$$

The quantity $[\,\Sigma\beta_i - \pi\,] = [\,2\pi - \Sigma\alpha_i\,]$ is also called the **excess** *of* Δ, and when the excess is negative, the positive quantity $[\,\pi - \Sigma\beta_i\,] = [\,\Sigma\alpha_i - 2\pi\,]$ is called the **defect** of Δ.

If you have not already seen it, note now the close connection between the holonomy, the excess, and the area of a triangle. Note that the holonomy is positive for triangles on a sphere and negative for triangles in a hyperbolic plane (and zero for triangles on a plane). One consequence of this formula is that the holonomy does not depend on either the vertex or the vector we start with. This is to be expected because parallel transport does not change the relative angles of any figure.

Following Problems **7.1**, **7.2**, and **7.4**, we can write the result for triangles on a sphere with radius ρ in this form:

Sphere:
$$\mathcal{H}(\Delta) = (\beta_1 + \beta_2 + \beta_3) - \pi = \text{Area}(\Delta)\ 4\pi/S_\rho = \text{Area}(\Delta)\ \rho^{-2}.$$

For a hyperbolic plane made with annuli with radius ρ, we get:

Hyperbolic:
$$\mathcal{H}(\Delta) = (\beta_1 + \beta_2 + \beta_3) - \pi = -\text{Area}(\Delta)\ \pi/I_\rho = -\text{Area}(\Delta)\ \rho^{-2}.$$

The quantity ρ^{-2} is traditionally called the **Gaussian curvature** or just plain **curvature** of the sphere and $-\rho^{-2}$ is called the (**Gaussian**) **curvature** of the hyperbolic plane. If K denotes the (*Gaussian*) curvature as just defined, then the formula

$$(\beta_1 + \beta_2 + \beta_3) - \pi = \text{Area}(\Delta)\ K$$

is called the ***Gauss-Bonnet Formula*** (for triangles). The formula is originally due to C. F. Gauss (1777–1855, German) and was extended by P. O. Bonnet (1819–1892, French), as we will describe at the end of this chapter.

Can you see how this result gives a bug on the surface an intrinsic way of determining the quantity K and thus also determining the extrinsic radius ρ?

The Gauss-Bonnet Formula not only holds for triangles in an open hemisphere or in a hyperbolic plane but can also be extended to any simple (that is, non-intersecting) polygon (that is, a closed curve made up of a finite number of geodesic segments) contained in an open hemisphere or in a hyperbolic plane.

*PROBLEM 7.5   GAUSS-BONNET FORMULA FOR POLYGONS

DEFINITION.   *The holonomy of a simple polygon,* $\mathcal{H}(\Gamma)$, *in an open hemisphere or in a hyperbolic plane is defined as follows:*

*If you parallel transport a vector (a directed geodesic segment) counterclockwise around the sides of the simple polygon, then the **holonomy** of the polygon is the smallest angle measured counterclockwise from the original position of the vector and its final position.*

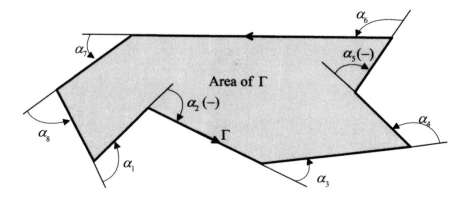

Figure 7.10   Exterior angles

If you walk around a polygon with the interior of the polygon on the left, the exterior angle at a vertex is the change in the direction at that vertex. This change is positive if you turn counterclockwise and negative if you turn clockwise. (See Figure 7.10.)

We will first look at convex polygons because this is the only case we will need later and it is easier to understand. A region is called **convex** if every pair of points in the region can be joined by a geodesic segment lying wholly in the region.

a. *Show that if Γ is a convex polygon in an open hemisphere or in a hyperbolic plane, then*

$$\mathcal{H}(\Gamma) = 2\pi - \Sigma\alpha_i = \Sigma\beta_i - (n-2)\pi = \text{Area}(\Gamma)\,K,$$

where $\Sigma\alpha_i$ is the sum of the exterior angles, $\Sigma\beta_i$ is the sum of the interior angles, n is the number of sides, and K is the Gaussian curvature.

Divide the convex polygon into triangles as in Figure 7.11. Now apply **7.4** to each triangle and carefully add up the results. You can check directly that $\mathcal{H}(\Gamma) = 2\pi - \Sigma\alpha_i$.

b. *Prove that every simple polygon on the plane or on a hemisphere or on a hyperbolic plane can be dissected into triangles without adding extra vertices.*

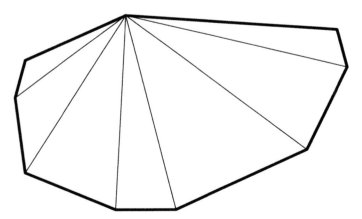

Figure 7.11   Dividing a convex polygon into triangles

SUGGESTIONS

Look at this on the plane, hemispheres, and hyperbolic planes. The difficulty in this problem is coming up with a method that works for all polygons, including very general or complex ones, such as the polygon in Figure 7.12.

Figure 7.12   General polygon

You may be tempted to try to connect nearby vertices to create triangles, but how do we know that this is always possible? How do you know that in any polygon there is even one pair of vertices that can be joined in the interior? The polygon may be so complex that parts of it get in the way of what you're trying to connect. So you might start by giving a convincing argument that there is at least one pair of vertices that can be joined by a segment in the interior of the polygon.

To see that there *is* something to prove here, there are examples of polyhedra in 3-space with **no** pair of vertices that can be joined in the interior. This interesting fact was first published in 1911 by N. J. Lennes; thus such polyhedra are often called *Lennes Polyhedra*. One example of a Lennes Polyhedron in depicted in Figure 7.13. The polyhedron consists of eight triangular faces and six vertices. Each vertex is joined by an edge to four of the other vertices, and the straight line segment joining it to the fifth vertex lies in the exterior of the polygon. Therefore, it is impossible to dissect this polyhedron into tetrahedra without adding extra vertices. This example and some history of the problem are discussed in [**DI**: Eves, p. 211] and [**DI**: Ho].

Note that there is at least one convex vertex (a vertex with interior angle less than π) on every polygon (in fact, it is not too hard to see that there must be at least three such vertices). To see this, pick any geodesic in the exterior of the polygon and parallel transport it toward the polygon until it first touches the polygon. It is easy to see that the line must now be intersecting the polygon at a convex vertex.

Figure 7.13   A polyhedron with vertices not joinable in the interior

    c. *Show that if Γ is a simple polygon in an open hemisphere or in a hyperbolic plane, then*

$$\mathcal{H}(\Gamma) = 2\pi - \Sigma\alpha_i = \Sigma\beta_i - (n-1)\pi = \text{Area}(\Gamma)\,K,$$

where $\Sigma\alpha_i$ is the sum of the exterior angles, $\Sigma\beta_i$ is the sum of the interior angles, and K is the Gaussian curvature.

Start by applying part **b.** Then proceed as in part **a**, but for this part you may find it easier to show that the holonomy of the polygon is the sum of the holonomies of the triangles by removing one triangle at a time. Again, you can check directly that $\mathcal{H}(\Gamma) = 2\pi - \Sigma\alpha_i$.

*GAUSS-BONNET FORMULA FOR POLYGONS ON SURFACES

The above discussion of holonomy is in the context of an open hemisphere and a hyperbolic plane, but the results have a much more general applicability and constitute an important aspect of differential geometry. In particular, we can extend this result even further to general surfaces, even those of non-constant curvature. In fact, Gauss defined the (***Gaussian***) ***curvature*** ***K(p)*** at a point p on any surface to be

$$K(p) = \lim_{\Delta \to p} \mathcal{H}(\Delta) \,/\, A(\Delta),$$

where the limit is taken over a sequence of small (geodesic) triangles that converge to p. The reader can check that the Gaussian curvature of a sphere (with radius ρ) is $1/\rho^2$ and that the Gaussian curvature of a hyperbolic plane (with radius ρ, the radius of the annular strips) is $-1/\rho^2$. This definition leads us to another formula, namely,

THEOREM 7.5a. *The Gauss-Bonnet Formula for Polygons on Surfaces*
On any smooth surface (2-manifold), if Γ is a (geodesic) polygon that bounds a contractible region, then

$$\mathcal{H}(\Gamma) = 2\pi - \Sigma \ \alpha_i \ = \iint_{I(\Gamma)} K(p) \ dA,$$

where the integral is the (surface) integral over $I(\Gamma)$, the interior of the polygon.

A region is said to be ***contractible*** if it can be continuously deformed to a point in its interior. See Figure 7.14 for examples.

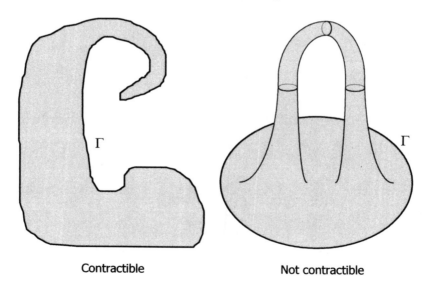

Contractible   Not contractible

Figure 7.14

The proof of this formula involves dividing the interior of Γ into many triangles, each so small that the curvature K is essentially constant over its interior, and then applying the Gauss-Bonnet Formula for spheres and hyperbolic planes to each of the triangles.

In addition, all of the versions of the Gauss-Bonnet Formula given thus far can be extended to arbitrary, simple, piecewise smooth, closed curves. (It is this extension that was Bonnet's contribution to the Gauss-Bonnet Formula.) If γ is such a curve, then we can define the holonomy $\mathcal{H}(\gamma) = \lim \mathcal{H}(\gamma_i)$, where the limit is over a sequence (which converges point-wise to γ) of geodesic polygons $\{\gamma_i\}$ whose vertices lie on γ. Using this definition, the Gauss-Bonnet Formula can be extended even further.

THEOREM 7.5b. *The Gauss-Bonnet Formula for Curves That Bound a Contractible Region*

On a sphere or hyperbolic plane, with (constant) curvature K,

$$\mathcal{H}(\gamma) = A(\gamma)\,K,$$

where $A(\gamma)$ is the area of the region bounded by γ.

On general surfaces,

$$\mathcal{H}(\gamma) = \iint_{I(\gamma)} K(p)\,dA,$$

where $I(\gamma)$ is the interior of the region bounded by γ.

Another version of the Gauss-Bonnet Formula is discussed in Problem **17.6**, where the integral is over the whole surface.

For further discussions, see *Differential Geometry: A Geometric Introduction* [**DG**: Henderson], Chapters 5 and 6, especially Problems **5.4** and **6.4**.

Chapter 8

PARALLEL TRANSPORT

Parallel straight lines are straight lines lying in a plane which do not meet if continued indefinitely in both directions.
— Euclid, *Elements*, Definition 23 [Appendix A]

In this chapter we will further develop the notion of *parallel transport* that was introduced in Chapter 7. This chapter may be studied independently of Chapter 7 if you read the section in Chapter 7 entitled Introducing Parallel Transport. The basic idea of Chapter 8 is to collect all the results related to parallelism that can be examined without assuming any special properties on the plane about parallel lines or about the sum of the angles on the plane. These properties (postulates) will be discussed in detail in Chapter 10.

PROBLEM 8.1 EUCLID'S EXTERIOR ANGLE THEOREM (EEAT)

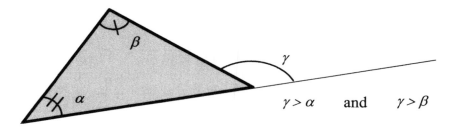

Figure 8.1   EEAT

a.   *Any exterior angle of a triangle is greater than each of the opposite interior angles.* **Warning**: Euclid's EAT is not the same as the Exterior Angle Theorem usually studied in high school.

b.   *When is EEAT true on the plane, on a sphere, and on a hyperbolic plane?*

SUGGESTIONS

You may find the following hint (which is found in Euclid's proof) useful: Draw a line from the vertex of α to the midpoint, M, of the opposite side, BC. Extend that line beyond M to a point A' in such a way that $AM \cong MA'$. Join A' to C. This hint will be referred to as *Euclid's hint* and is pictured in Figure 8.2.

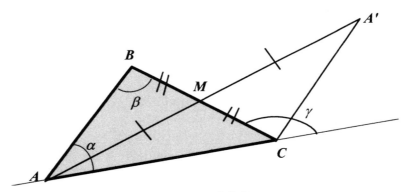

Figure 8.2   Euclid's hint

Be cautious transferring this hint to a sphere. It will probably help to draw Euclid's hint directly on a physical sphere.

It is not necessary to use Euclid's hint to prove EEAT, and in fact many people don't "see" the hint. Another perfectly good way to prove EEAT is to use Problem **8.2**. Problems **8.1** and **8.2** are very closely related, and they can be done in either order. It is also fine to use **8.1** to prove **8.2** or use **8.2** to prove **8.1**, but of course don't do both. As a final note, remember you do not have to look at figures using only one orientation — rotations and reflections of a figure do not change its properties, so if you have trouble "seeing" something, check to see if it's something you're familiar with by orienting it differently on the page.

EEAT is not always true on a sphere, even for small triangles. Look at a counterexample as depicted in Figure 8.3. Then look at your proof of

EEAT on the plane. It is very likely that your proof uses properties of angles and triangles that are true for small triangles on the sphere. Thus it may appear to you that your planar proof is also a valid proof of EEAT for small triangles on the sphere. But there is a counterexample.

This could be, potentially, a very creative situation for you — **whenever you have a proof and counterexample of the same result, you have an opportunity to learn something deep and meaningful**. So try out your planar proof of EEAT on the counterexample in Figure 8.3 and see what happens. Then try it on both large and small spherical triangles. If you can determine exactly which triangles satisfy EEAT and which triangles don't satisfy EEAT, then this information will be useful (but not crucial) to you in later problems.

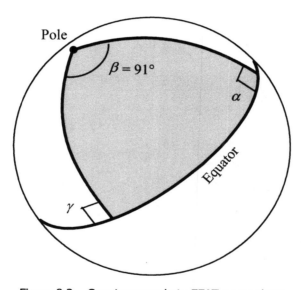

Figure 8.3   Counterexample to EEAT on a sphere

PROBLEM 8.2 SYMMETRIES OF PARALLEL TRANSPORTED LINES

*Consider two lines, **r** and **r'**, that are parallel transports of each other along a third line, **l**. Consider now the geometric figure that is formed by the three lines and look for the symmetries of that geometric figure. See Figure 8.4.*

*What can you say about the lines **r** and **r'**? Do they intersect? If so, where? Look at the plane, spheres, and hyperbolic planes.*

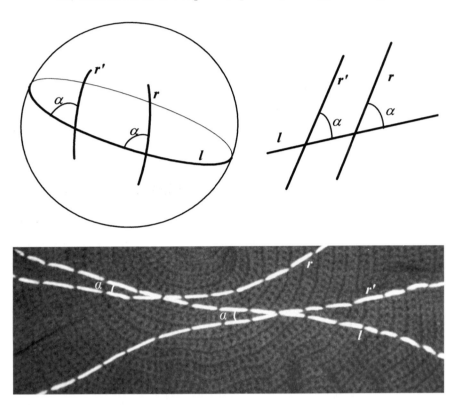

Figure 8.4   What can you say about **r** and **r'**?

SUGGESTIONS

Parallel transport was already *informally* introduced in Chapter 7. In Problem **8.2** you have an opportunity to explore the concept further and prove its implications on the plane and a sphere. You will study the relationship between parallel transport and parallelism, as well.

It is common in high school to use Euclid's definition of parallel lines as "straight lines lying in a plane which do not meet if continued indefinitely in both directions." But this is an inhuman definition — there is no way to check all points on both lines to see if they ever meet. This definition is also irrelevant on a sphere because we know that all geodesics on a sphere *will* cross each other. But we *can* measure the angles of a transversal. This is why it is more useful to talk about lines as parallel transports of one another rather than as parallel. So the question becomes

> *If a transversal cuts two lines at congruent angles, are the lines in fact parallel in the sense of not intersecting?*

There are many ways to approach this problem. First, be sure to look at the symmetries of the local portion of the figure formed by the three lines. See what you can say about global symmetries from what you find locally. For the question of parallelism, you can use EEAT, but not if you used this problem to prove EEAT previously. Also, don't underestimate the power of symmetry when considering this problem. Many ideas that work on the plane will also be useful on a sphere and a hyperbolic plane, so try your planar proof on a sphere and a hyperbolic plane before attempting something completely different.

In Chapter 1, we said that an isometry is a *symmetry* of a geometric figure if it transforms that figure into itself. That is, the figure looks the same before and after the isometry. Here, we are looking for the symmetries of the figure on the plane and sphere and hyperbolic plane.

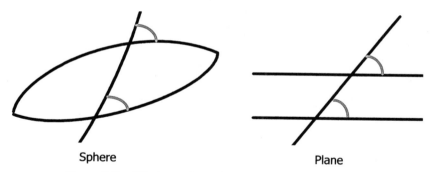

Sphere   Plane

Figure 8.5   What are the symmetries of these figures?

From Figure 8.5, we can see that on a sphere we are looking for the symmetries of a *lune* cut at congruent angles by a geodesic. A *lune* is a spherical region bounded by two half great circles (see Problem **7.1**).

You may be inclined to use one or both of the following results that are true on the plane: *Any transversal of a pair of parallel lines cuts these lines at congruent angles* (Problem **10.1**). And, *the angles of any triangle add up to a straight angle* (Problems **7.3b** and **10.2**). The use of these results should be avoided for now, as they are both false on both a sphere and a hyperbolic plane. We have been investigating what is common between the plane, spheres, hyperbolic planes — trying to use common proofs whenever possible. You may be tempted to use other properties of parallel lines that seem familiar to you, but in each case ask yourself whether or not the property is true on a sphere and on a hyperbolic plane. If it is not true on these surfaces, then don't use it here because it is not needed.

PROBLEM 8.3   TRANSVERSALS THROUGH A MIDPOINT

a. Prove: *If two geodesics r and r' are parallel transports along another geodesic l, then they are also parallel transports along any geodesic passing through the midpoint of the segment of l between r and r'. Does this hold for the plane, spheres, and hyperbolic planes?* See Figure 8.6.

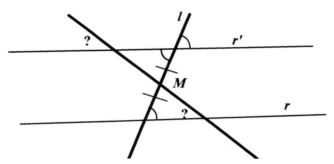

Figure 8.6 Transversals through a midpoint

b. *On a sphere or hyperbolic plane, are these the only lines that will cut r and r' at congruent angles? Why?*

c. Prove: *Two geodesics (on the plane, spheres, or hyperbolic planes) are parallel transports of each other if and only if they have a common perpendicular.*
    Is there only one common perpendicular?

All parts of this problem continue the ideas presented in Problem **8.2**. In fact, you may have proven this problem while working on **8.2** without even knowing it. There are many ways to approach this problem. Using symmetry, is always a good way to start. You can also use some of the triangle congruence theorems that you have been working with in Chapter 6. Look at the things you have discovered about transversals from Problems **8.1** and **8.2**; they are very applicable here. For the hyperbolic plane you may want to use results from Chapters 5 and/or 7.

PROBLEM 8.4   WHAT IS "PARALLEL"?

Since Chapter 7, you have been dealing with issues of parallelism. Parallel transport gives you a way to check parallelism. Even though parallel transported lines intersect on the sphere, there is a *feeling of local parallelness* about them. In most applications of parallel lines the issue is not whether the lines ever intersect, but whether a transversal intersects them at congruent angles at certain points; that is, whether the lines are *parallel transports of each other along the transversal*. You may choose to avoid definitions of "parallel" that do not give you a direct method of verification, such as these common definitions for parallel lines in the plane:

1. *Parallel lines are lines that never intersect;*

2. *Parallel lines are lines such that any transversal cuts them at congruent angles*; or

3. *Parallel lines are lines that are everywhere equidistant.*

   a. *Check out for each of these three definitions whether they apply to parallel transported lines on a sphere or on a hyperbolic plane.*

This is closely related to Problems **8.2** and **8.3**.

   b. *Show that there are pairs of geodesics on a hyperbolic plane that do not intersect and yet there are **no** transversals that cut at congruent angles. That is, the geodesics are parallel (in the sense of not intersecting) but not parallel transports of each other.*

Use the results of **8.2** and **8.3** and look on your hyperbolic plane for the boundary between geodesics that intersect and geodesics that are parallel transports. We call a pair of geodesics that satisfy part **b** by the name

asymptotic geodesics. Be warned that in many texts these geodesics are called simply *parallel*.

> ***c.** *Show that there are pairs of geodesics on any cone with cone angle greater than 360° that do not intersect and yet there are* ***no*** *transversals that cut at congruent angles. That is, the geodesics are parallel (in the sense of not intersecting) but not parallel transports of each other along any straight (in the sense of symmetry) line.*

Experiment with a paper cone with cone angle greater than 360°.

These examples should help you realize that parallelism is not just about non-intersecting lines and that the meaning of parallel is different on different surfaces. You will explore and discuss these various notions of parallelism and parallel transport further in Chapters 9–13. Because we have so many (often unconscious) connotations and assumptions attached to the word "parallel," we find it best to avoid using the term "parallel" as much as possible in this discussion. Instead we will use terms such as "parallel transport," "non-intersecting," and "equidistant," which make explicit the meaning that is intended.

In Chapter 9, we will continue our explorations of triangle congruence theorems, some of which involve parallel transport.

In Chapter 10, we will consider various parallel postulates and explore how they apply on the plane, spheres, and hyperbolic planes. We will assume the parallel postulates on the plane and use them to prove the properties of non-intersecting and parallel transported lines on the plane. In the process, you may learn something about the history and philosophy of parallel lines and the postulates that have been used in attempts to understand parallelism.

In Chapter 11, our understanding of different notions of "parallel" will help us to explore isometries and patterns.

In Chapter 12, we will study parallelograms and rectangles (and their analogues on spheres and hyperbolic planes) and in the process show that, on the plane, non-intersecting lines are equidistant.

In Chapter 13, we will use the results from Chapter 12 to explore results that are only true on the plane, such as the Pythagorean Theorem and results about similar triangles.

Chapter 9

SSS, ASS, SAA, AND AAA

> Things which coincide with one another are equal to one another.
> — Euclid, *Elements*, Common Notion 4 [Appendix A]

This chapter is a continuation of the triangle congruence properties studied in Chapter 6.

PROBLEM 9.1   SIDE-SIDE-SIDE (SSS)

Are two triangles congruent if the two triangles have congruent corresponding sides? Look at plane, spheres, and hyperbolic planes. See Figure 9.1.

Figure 9.1   SSS

SUGGESTIONS

Start investigating SSS by making two triangles coincide as much as possible, and see what happens. For example, in Figure 9.2, if we line up one pair of corresponding sides of the triangles, we have two different orientations for the other pairs of sides as depicted in Figure 9.2. Of course, it is up to you to determine if each of these orientations is actually possible, and to prove or disprove SSS. Again, symmetry can be very useful here.

117

Figure 9.2   Are these possible?

On a sphere, SSS doesn't work for all triangles. The counterexample in Figure 9.3 shows that no matter how small the sides of the triangle are, SSS does not hold because the three sides always determine two different triangles on a sphere. Thus, it is necessary to restrict the size of more than just the sides in order for SSS to hold on a sphere. Whatever argument you used for the plane should work for *suitably defined* small triangles on the sphere and all triangles on a hyperbolic plane. Make sure you see what it is in your argument that doesn't work for large triangles on a sphere.

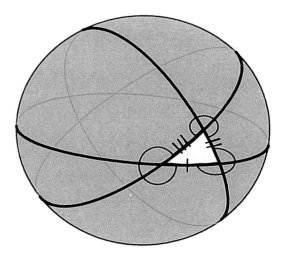

Figure 9.3   A large triangle with small sides

There are also other types of counterexamples to SSS on a sphere. Can you find them?

PROBLEM 9.2 ANGLE-SIDE-SIDE (ASS)

a. *Are two triangles congruent if an angle, an adjacent side, and the opposite side of one triangle are congruent to an angle, an adjacent side, and the opposite side of the other? Look at plane, spheres, and hyperbolic planes.* See Figure 9.4.

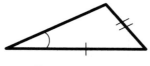

Figure 9.4   ASS

SUGGESTIONS

Suppose you have two triangles with the above congruencies. We will call them ASS triangles. We would like to see if, in fact, the triangles are congruent. We can line up the angle and the first side, and we know the length of the second side (*BC* or *B'C'*), but we don't know where the second and third sides will meet. See Figure 9.5.

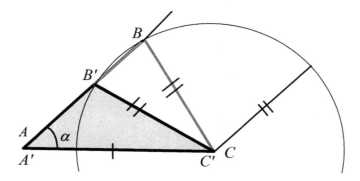

Figure 9.5   ASS is not true, in general

Here, the circle that has as its radius the second side of the triangle intersects the ray that goes from *A* along the angle α to *B* twice. So ASS doesn't work for all triangles on the plane or spheres or hyperbolic planes. Try this for yourself on these surfaces to see what happens. Can you make ASS work for an appropriately restricted class of triangles? On a sphere, also look at triangles with multiple right angles, and, again, define "small triangles" as necessary. Your definition of "small triangle" here may be very different from your definitions in Problems **6.4** and **6.5**.

There are numerous collections of triangles for which ASS is true. Explore. See what you find on all three surfaces.

> **b.**   *Show that ASS holds for right triangles on the plane (where the Angle in Angle-Side-Side is right).*

This result is often called the *Right-Leg-Hypotenuse Theorem* (RLH), which can be expressed in the following way:

> **RLH:**   *On the plane, if the leg and hypotenuse of one right triangle are congruent to the leg and hypotenuse of another right triangle, then the triangles are congruent.*

> *What happens on a sphere and a hyperbolic plane?*

At this point, you might conclude that RLH is true for small triangles on a sphere. But there *are* small triangle counterexamples to RLH on spheres! The counterexample in Figure 9.6 will help you to see some ways in which spheres are intrinsically very different from the plane. We can see that the second leg of the triangle intersects the geodesic that contains the third side an infinite number of times. So on a sphere there are small triangles that satisfy the conditions of RLH although they are not congruent. What about RLH on a hyperbolic plane?

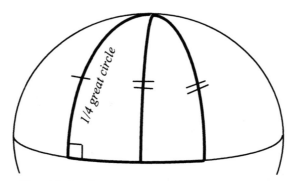

Figure 9.6   Counterexample to RLH on a sphere

However, if you look at your argument for RLH on the plane, you should be able to show the following:

On a sphere, RLH is valid for a triangle with all sides less than 1/4 of a great circle.

RLH is also true for a much larger collection of triangles on a sphere. Can you find such a collection? What about on a hyperbolic plane?

PROBLEM **9.3** SIDE-ANGLE-ANGLE (SAA)

Are two triangles congruent if one side, an adjacent angle, and the opposite angle of one triangle are congruent, respectively, to one side, an adjacent angle, and the opposite angle of the other triangle? Look at plane, spheres, and hyperbolic planes.

SUGGESTIONS

As a general strategy when investigating these problems, start by making the two triangles coincide as much as possible. You did this when investigating SSS and ASS. Let us try it as an initial step in our proof of SAA. Line up the first sides and the first angles. Because we don't know the length of the second side, we might end up with a picture similar to Figure 9.7.

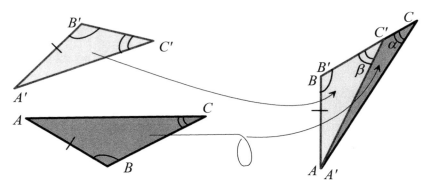

Figure 9.7   Starting SAA

The situation shown in Figure 9.7 may seem to you to be impossible. You may be asking yourself, "Can this happen?" If your temptation is to argue that α and β cannot be congruent angles and that it is not possible to construct such a figure, behold Figure 9.8.

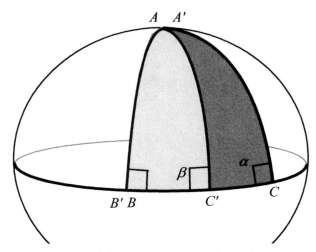

Figure 9.8   A counterexample to SAA

You may be suspicious of this example because it is not a counter-example on the plane. You may feel certain that it is the only counterex-ample to SAA on a sphere. In fact, we can find other counterexamples for SAA on a sphere.

With the help of parallel transport, you can construct many counter-examples for SAA on a sphere. If you look back to the first counterex-ample given for SAA, you can see how this problem involves parallel transport, or similarly how it involves Euclid's Exterior Angle Theorem, which we looked at in Problem **8.1**.

Can we make restrictions such that SAA *is* true on a sphere? You should be able to answer this question by using the fuller understanding of parallel transport you gained in Problems **8.1** and **8.2**. You may be tempted to use the result, *the sum of the interior angles of a triangle is 180°*, in order to prove SAA on the plane. This result will be proven later (Problem **10.2**) for the plane, but we saw in Problem **7.3** that it does not hold on spheres and hyperbolic planes. Thus, we encourage you to avoid using it and to use the concept of parallel transport instead. This suggestion stems from our desire to see what is common between the plane and the other two surfaces, as much as possible. In addition, before we can prove that the sum of the angles of a triangle is 180°, we will have to make some additional assumptions on the plane that are not needed for SAA.

PROBLEM **9.4**   ANGLE-ANGLE-ANGLE (AAA)

Are two triangles congruent if their corresponding angles are congruent? Look at plane, spheres, and hyperbolic planes.

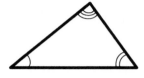

Figure 9.9   AAA

Two triangles that have corresponding angles congruent are called *similar triangles*. We will discuss similar triangles in Problem **13.4**.

As with the three previous problems, make the two AAA triangles coincide as much as possible. We know that we can line up one of the angles, but we don't know the lengths of either of the sides coming from this angle. So there are two possibilities: (1) Both sides of one are longer than both sides of the other, as the example in Figure 9.10 shows on the plane, or (2) one side of the first triangle is longer than the corresponding side of the second triangle and vice versa, as the example in Figure 9.11 shows on a sphere.

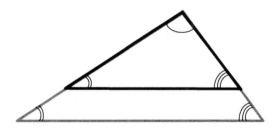

Figure 9.10   Is this possible?

As with Problem **9.3**, you may think that the example in Figure 9.11 cannot happen on a plane, a sphere, or a hyperbolic plane. The possible existence of a counterexample relies heavily on parallel transport — you can identify the parallel transports in each of the examples given. Try each counterexample on the plane, on a sphere, and on a hyperbolic plane and see what happens. If these examples are not possible, explain why, and if they are possible, see if you can restrict the triangles sufficiently so that AAA does hold.

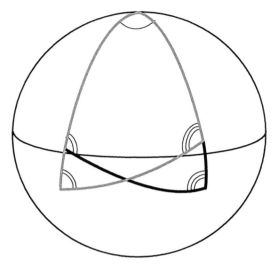

Figure 9.11   Is this possible?

Parallel transport shows up in AAA, similar to how it did in SAA, but here it happens simultaneously in two places. In this case, you will recognize that parallel transport produces similar triangles that are not necessarily congruent. Are these constructions (in Figures 9.10 and 9.11) possible? How? Are the triangles not congruent? Why? These constructions and non-congruencies may seem intuitively possible to you, but you should justify why in each case. Again you may need properties of angle sums from Problem **7.3** and properties of parallel transport from Problems **8.1** through **8.4**. You may also use the property of parallel transport on the plane stated in Problem **10.1** — you can assume this property now as long as you are sure not to use AAA when proving it later.

On a sphere or on a hyperbolic plane, is it possible to make the two parallel transport constructions shown in Figure 9.11 and thus get two non-congruent triangles? Try it and see. It is important that you make such constructions and that you study them on a model of a sphere and on a model of a hyperbolic plane.

Chapter 10

PARALLEL POSTULATES

[Euclid's Fifth Postulate] ought to be struck from the postulates altogether. For it is a theorem — one that invites many questions ... — and requires for its demonstration a number of definitions as well as theorems. ... it lacks the special character of a postulate.

— Proclus (Greek, 410–485) [**AT**: Proclus], p. 151

PARALLEL LINES ON THE PLANE ARE SPECIAL

Up to this point we have not had to assume anything about parallel (non-intersecting) lines. No version of a parallel postulate has been necessary, on the plane, on a sphere, or on a hyperbolic plane. We defined the concrete notion of parallel transport and proved in Problem **8.2** that, on the plane (and hyperbolic planes), parallel transported lines do not intersect. Now in this chapter we will look at three important properties on the plane that require further assumptions and that will be needed in later chapters. If you are willing to assume these three statements, you may skip this chapter, but we urge you to finish reading through at least the next page.

If two lines on the plane are parallel transports of each other along some transversal, then they are parallel transports along any transversal. (Contained in Problem **10.1**)

On the plane, the sum of the interior angles of a triangle is always 180°. (Contained in Problem **7.3b** or Problem **10.2**)

On the plane, non-intersecting lines are parallel transports along all transversals. (Contained in Problem **10.3d**)

We have already seen that none of these properties is true on a sphere or a hyperbolic plane. Thus all three need for their proofs some property of the plane that does not hold on spheres and hyperbolic planes. The various properties that permit proofs of these three statements are collectively termed the ***Parallel Postulates***.

Only the three statements above are needed from this chapter for the rest of the book. Therefore, it is possible to omit this chapter and assume one of the above three statements and then prove the others. However, parallel postulates have a historical importance and a central position in many geometry textbooks and in many expositions about non-Euclidean geometries. The problems in this chapter are an attempt to help people unravel and enhance their understanding of parallel postulates. Comparing situations on the plane with situations on a sphere and on a hyperbolic plane is a powerful tool for unearthing our hidden assumptions and misconceptions about the notion of "parallel" on the plane.

As we discussed already in Chapter 8, because we have so many (often unconscious) connotations and assumptions attached to the word "parallel," we find it best to avoid using the term "parallel" as much as possible in this discussion. Instead we will use terms such as "parallel transport," "non-intersecting," and "equidistant," which make explicit the meaning that is intended.

PROBLEM 10.1   PARALLEL TRANSPORT ON THE PLANE

Show that if l_1 and l_2 are lines on the plane such that they are parallel transports along a transversal l, then they are parallel transports along any transversal. Prove this using any assumptions you find necessary. Make as few assumptions as you can, and make them as simple as possible. ***Be sure to state your assumptions clearly.***

What part of your proof does not work on a sphere or on a hyperbolic plane?

SUGGESTIONS

This problem is by no means as trivial as it at first may appear. In order to prove this theorem, you will have to assume something — there are many possible assumptions, so use your imagination. But at the same time, try not to assume any more than is necessary. If you are having trouble deciding what to assume, try to solve the problem in a way that seems natural to you and then see what develops while making explicit any assumptions you are using.

On spheres and hyperbolic planes, try the same construction and proof you used for the plane. What happens? You should find that your proof does not work on these surfaces. So what is it about your proof (on a sphere and hyperbolic plane) that creates difficulties?

Problem **10.1** emphasizes the differences between parallelism on the plane and parallelism on spheres and hyperbolic planes. On the plane, non-intersecting lines exist, and one can "parallel transport" everywhere. Yet, as was seen in Problems **8.2** and **8.3**, on spheres and hyperbolic planes two lines are cut at congruent angles if and only if the transversal line goes through the center of symmetry formed by the two lines. That is, on spheres and hyperbolic planes two lines are parallel transports only when they can be parallel transported through the center of symmetry formed by them. Be sure to draw a picture locating the center of symmetry and the transversal. On spheres and hyperbolic planes it is impossible to slide the transversal along two parallel transported lines while keeping both angles constant (something you *can* do on the plane). In Figures 10.1, the line r' is a parallel transport of line r along line l, but it is not a parallel transport of r along l'.

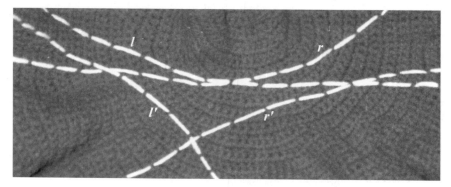

Figure 10.1a   Parallel transport on a hyperbolic plane along *l* but not along *l'*

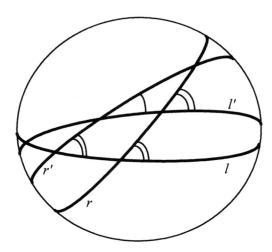

Figure 10.1b   Parallel transport on a sphere along *l*, but not along *l'*

Pause, explore, and write out your ideas before reading further.

We will now divide parallel postulates into three groups: those involving mostly parallel transport, those involving mostly equidistance, and those involving mostly intersecting or non-intersecting lines. This division is useful even though it is rough and unlikely to fit every conceivable parallel postulate. *Which of these three groups is the most appropriate for your assumption from Problem* **10.1**? We will call the assumption you made in Problem **10.1** *"your parallel postulate."*

PROBLEM 10.2   PARALLEL POSTULATES NOT INVOLVING (NON-)INTERSECTING LINES

Commonly used postulates of this sort are

H = 0:   *The holonomy of triangles is zero.*

A = 180:   *The sum of the angles of a triangle is equal to 180°.*

> **PT!**:   *If two lines are parallel transports (PT) along one line*
> *then they are PT along ALL transversals.*

Note that these are false on spheres and on hyperbolic planes. The last two properties (**A** = 180 and **PT!**) will be needed crucially in almost all the remaining chapters (and you may have already used one of them in Problem **9.4**). These properties are needed to study (on the plane) parallelograms, rectangles, and similar triangles (Chapters 12 and 13), circles and inversion (Chapters 15 and 16), projections of spheres and hyperbolic planes onto the Euclidean plane (Chapters 14 and 17), Euclidean manifolds (Chapters 18 and 24), solutions to quadratic and cubic equations (Chapter 19), trigonometry (Chapter 20), and polyhedra in 3-space (Chapter 23). It is these properties and their uses that we most often associate with consequences of the parallel postulates.

In Problem **7.3b** we proved **A** = 180 on the plane, so we can

a.   *Ask what assumption about the plane was used in proving* **7.3b**.

b.   *Use* **A** = 180 *to prove* **PT!**.

Look first at transversals that intersect the line along which the two lines are parallel transports.

c.   *Using* **PT!** *(Problem* **10.1**)*, prove* **A** = 180 *without using results from Chapter 7.*

Start by parallel transporting one side of the triangle.

d.   *Show that, on the plane,* **H** = **0** ⟺ **A** = 180 ⟺ **PT!**. *If your postulate from* **10.1** *is in this group, then show that it is also equivalent to the others.*

This is usually accomplished most efficiently by proving **H** = **0** ⟹ **A** = 180 ⟹ **PT!** ⟹ Your Postulate ⟹ **H** = **0**, or in any other order.

e.   *Prove that, on the plane, two parallel transported lines are equidistant.*

Look for rectangles or parallelograms.

EQUIDISTANT CURVES ON SPHERES
AND HYPERBOLIC PLANES

The latitude circles on the earth are sometimes called "parallels of latitude." They are parallel in the sense that they are everywhere equidistant as are concentric circles on the plane. In general, transversals do not cut equidistant circles at congruent angles. However, there is one important case where transversals do cut the circles at congruent angles. Let *l* and *l'* be latitude circles the same distance from the equator on opposite sides of it. See Figure 10.2. Then every point on the equator is a center of half-turn symmetry for these pair of latitudes. Thus, as in Problems **8.3** and **10.1**, every transversal cuts these latitude circles in congruent angles, even though these latitude circles are not geodesics. In the first section of Chapter 2 we noted that Euclid, in his *Phenomena*, discussed such equidistant circles.

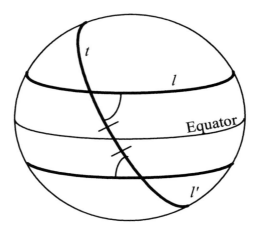

Figure 10.2   Special equidistant circles

The same ideas work on a hyperbolic plane: If *g* is a geodesic and *l* and *l'* are the (two) curves (not geodesics) that are a distance *d* from *g*, then *l* and *l'* are equidistant from each other and every transversal cuts them at congruent angles. This follows from the fact that *g* has half-turn symmetry at every point.

PROBLEM **10.3** PARALLEL POSTULATES INVOLVING (NON-)INTERSECTING LINES

One of Euclid's assumptions constitutes *Euclid's Fifth (or Parallel) Postulate (EFP)*, which says

> **EFP:** *If a straight line intersecting two straight lines makes the interior angles on the same side less than two right angles, then the two lines (if extended indefinitely) will meet on that side on which are the angles less than two right angles.*

For a picture of EFP, see Figure 10.3.

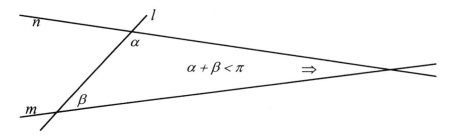

Figure 10.3   Euclid's Parallel Postulate

You probably did not assume EFP in your proof of Problem **10.1**. You are in good company — many mathematicians, including Euclid, have tried to avoid using it as much as possible. However, we will explore EFP because, historically, it is important, and because it has some very interesting properties, as you will see in Problem **10.3**. On a sphere, all straight lines intersect twice, which means that EFP is trivially true on a sphere. But in Problem **10.4**, you will show that EFP is also true in a stronger sense on spheres. You will also be able to prove that EFP is false on a hyperbolic plane.

Thus, EFP does not have to be assumed on a sphere — it can be proved! However, in most high school geometry textbooks, EFP is replaced by another postulate, claimed to be equivalent to EFP. This postulate we will call the ***High School Parallel Postulate*** (***HSP***), and it can be expressed in the following way:

HSP: *For every line **l** and every point **P** not on **l**, there is a unique line **l'** that passes through **P** and does not intersect (is parallel to) **l**.*

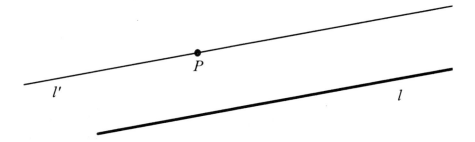

Figure 10.4   High School Parallel Postulate

In many books the **HSP** is called "Playfair's Parallel Postulate", but this is an inappropriate name, as we will show in the historical notes in the last two sections of this chapter.

Note that, because any two great circles on a sphere intersect, there are *no* lines *l'* that are parallel to *l* in the "not intersecting" sense. Therefore, **HSP** is not true on spheres. On the other hand, if we change "parallel" to "parallel transport" then *every* great circle through *P* is a parallel transport of *l* along *some* transversal. What happens on a hyperbolic plane? In Problem **10.3**, you will explore the relationships among **EFP**, **HSP**, and your postulate from Problem **10.1**. In Problem **10.4** we will explore these postulates on spheres and hyperbolic planes.

a. *Show that, on the plane, **EFP** and **HSP** are equivalent. If your postulate from **10.1** is in this group, is it equivalent to the others? Why or why not?*

To show that **EFP** and **HSP** are equivalent on the plane, you need to show that you can prove **EFP** if you assume **HSP** and vice versa. If the three postulates are equivalent, then you can prove the equivalence by showing that

$$\textbf{EFP} \Rightarrow \textbf{HSP} \Rightarrow \text{Your Postulate} \Rightarrow \textbf{EFP}$$

or in any other order. It will probably help you to draw lots of pictures of what is going on. Also, remember that we proved in Problem **8.2** (without using any parallel postulate) that parallel (non-intersecting) lines exist. Note that **HSP** is not true on a sphere but **EFP** is true, so your proof that **EFP** implies **HSP** on the plane must use some property of the plane that does not hold on a sphere. Look for it.

> **b.** *Prove that either **EFP** or **HSP** can be used to prove (without using **10.1** or **7.3b**) one of **H = 0**, **A = 180**, or **PT!**. (It does not matter which. Why?)*

So, are all these postulates equivalent to each other on the plane? The answer is almost, but not quite! In order for **A = 180** (or **H = 0** or **PT!**) to imply **EFP** (or **HSP**), we have to make an additional assumption, the ***Archimedean Postulate*** (**AP**) (in some books this is called the *Axiom of Continuity*), named after the Greek mathematician, Archimedes (who lived in Sicily, 287?–212 B.C.):

> ***AP**: *On a line, if the segment AB is less than* (contained in) *the segment AC, then there is a finite* (positive) *integer, n, such that, if we put n copies of AB end-to-end (see Figure 10.5), then the nth copy will contain the point C.*

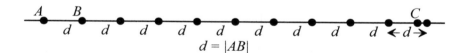

$$d = |AB|$$

Figure 10.5   The Archimedean Postulate

The Archimedean Postulate can also be interpreted to rule out the existence of infinitesimal lengths. The reason these postulates are "almost but not quite" equivalent to each other on the plane is that, though **AP** is needed, it is assumed by most people to be true on the plane, spheres, and hyperbolic planes. But this is the first time we have needed this assumption in this book.

> ***c.** *Show that, on the plane, **AP** and **A = 180** imply **EFP**.*

Look at the situation of **EFP**, which we redraw in Figure 10.6. Pick a sequence of equally spaced points A, A', A'',... on n (on the side of the angle α). Next, parallel transport l along m to lines l', l'',... that intersect n at the points A', A'',.... And parallel transport m along l to m', m'',..., which also intersect n at the points A', A'',.... Look for congruent angles and congruent triangles. Use **AP** to argue that n and m will eventually intersect.

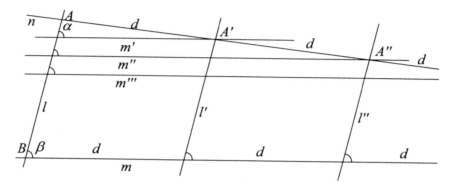

Figure 10.6   **AP** and **A = 180** imply **EFP**

Clearly, this proof feels different from the others proofs in this chapter. And, as already pointed out, most of the applications in this book of parallel postulates only need **A = 180** and **PT!**. So is it possible for us not to bother with **EFP** and **HSP**? Or are they needed? Yes, we need them, but only in a few places, such as in Problem **11.3**, where what we need to know is

> **d.** Prove using **EFP**: *On the plane, non-intersecting lines are parallel transports along every transversal.*

Compare with **EFP**.

PROBLEM 10.4   EFP AND HSP ON SPHERE
AND HYPERBOLIC PLANE

> **a.** *Show that **EFP** is true on a sphere in a strong sense; that is, if lines **l** and **l'** are cut by a transversal **t** such that the sum of the interior angles $\alpha + \beta$ on one side is less than two right angles, then not only do **l** and **l'** intersect, but they also intersect closest*

*to **t** on the side of **α** and **β**. You will have to determine an appropriate meaning for "closest."*

To help visualize the postulates, draw these "parallels" on an actual sphere. There are really two parts in this proof — first, you must come up with a definition of "closest," and then prove that **EFP** is true for this definition. The two parts may come about simultaneously as you come up with a proof. This problem is closely related to Euclid's Exterior Angle Theorem but can also be proved without using **EEAT**. One case that you should look at specifically is pictured in Figure 10.7. It is not necessarily obvious how to define the "closest" intersection.

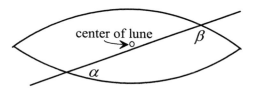

Figure 10.7   Is **EFP** true on a sphere?

b. *On a hyperbolic plane let **l** be a geodesic and let **P** be a point not on **l**; then show that there is an angle θ with the property that any line **l'** passing through **P** is parallel to (not intersecting) **l** if the line **l'** does not form an angle less than θ with the line from **P** that is perpendicular to **l** (Figure 10.8).*

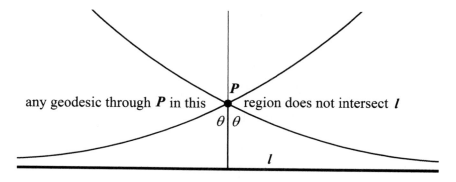

Figure 10.8   Multiple parallels on a hyperbolic plane

Look at a variable line through *P* that does intersect *l* and then look at what happens as you move the intersection point out to infinity.

> **c.** *Using the notion of parallel transport, change the High School Postulate so that the changed postulate is true on both spheres and hyperbolic planes. Make as few alterations as possible and keep some form of uniqueness.*

Try to limit your alterations so that the new postulate preserves the spirit of the old one. You can draw ideas from any of the previous problems to obtain suitable modifications. Prove that your modified versions of the postulate are true on spheres and hyperbolic planes.

> **d.** *Either prove your postulate from Problem **10.1** on a sphere and on a hyperbolic plane or change it, with as few alterations as possible, so that it is true on these surfaces. You may need to make different changes for the two surfaces.*

In Problem **10.1** you should have decided whether or not your postulate is true on spheres or on hyperbolic spaces.

Comparisons of Plane, Spheres, and Hyperbolic Planes

Figure 10.9 is an attempt to represent the relationships among parallel transport, non-intersecting lines, **EFP**, and **HSP**. Can you fit your postulate into the diagram?

The High School Postulate (**HSP**) assumes both the existence and uniqueness of parallel lines. In Problem **8.2**, you proved that if one line is a parallel transport of another, then the lines do not intersect on the plane or on a hyperbolic plane. Thus, it is not necessary to assume the existence of parallel lines (non-intersecting lines) on the plane or hyperbolic plane.

On a sphere any two lines intersect. However, in Problem **8.4** we saw that there *are* non-intersecting lines that are not parallel transports of each other on a hyperbolic plane and on any cone with cone angle larger than 360°.

	Euclidean	Hyperbolic	Spherical
Parallel transported lines	do not intersect and are equidistant **8.2 & 10.2**	diverge in both directions **8.2 & 8.3**	always intersect **2.1**
Parallel transported lines	are parallel transports along all transversals **10.1**	are parallel transports along any transversal that passes through the center of symmetry of the lines **8.3**	
Non-intersecting lines	are parallel transports along all transversals **10.3d**	sometimes are not parallel transports **8.4b**	do not exist **2.1**
Euclid's 5th Postulate	must be assumed Chapter 10	is false **8.4b**	is provable in a strong sense **10.4a**
High School Parallel Postulate	a unique line through a point not intersecting a given line **10.3**	many lines through a point not intersecting a given line **10.4b**	no non-intersecting lines **2.1**
Your postulate from **10.1**			
Two points determine	a unique line and line segment **6.1d**		at least two line segments **6.1d**
Sum of the angles of a triangle	= 180° **7.3b & 10.2**	< 180° **7.1**	> 180° **7.2**
Holonomy	= 0° **10.2**	< 0° **7.4**	> 0° **7.4**
VAT & ITT	are always true, **3.2 & 6.2**		
SAS, ASA, and SSS	hold for all triangles **6.4, 6.5, & 9.1**		hold for small triangles **6.4, 6.5, & 9.1**
AAA	is false, **9.4**, similar triangles	is true, **9.4**, no similar triangles	

Figure 10.9   Comparisons of the three geometries

PARALLEL POSTULATES WITHIN THE
BUILDING STRUCTURES STRAND

The problem of parallels puzzled Greek geometers (see the quote at the beginning of this chapter). The prevailing belief was that it surely follows from the straightness of lines that lines like *n* and *m* in Figure 10.10 *had* to intersect, and so Euclid's Fifth Postulate would turn out to be unnecessary.

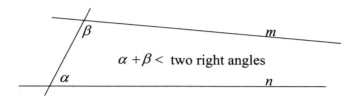

Figure 10.10   Surely these intersect!

During and since the Greek era, there have been attempts to derive the postulate from the rest of elementary geometry; attempts to reformulate the postulate or the definition of parallels into something less objectionable; and descriptions of what geometry would be like if the postulate was in some way denied. Part of these attempts was to study what geometric properties could be proved without using a parallel postulate, but with the other postulates (properties) of the plane. The results of these studies became known as ***absolute geometry***. Both the plane and a hyperbolic plane are examples of absolute geometry. Included in absolute geometry are all the results about the plane and hyperbolic planes that we have discussed in Chapters 1, 3, 5, 6, 8, and 9 (except AAA, Problem **9.4**). In general, absolute geometry includes everything that is true of both the plane and hyperbolic planes.

Claudius Ptolemy (Greek, 100–178) implicitly used what we now call Playfair's Postulate (**PP**) to prove **EFP**. This is reported to us by Proclus (Greek, 410–485) [**AT:** Proclus], who explicitly states **PP**.

> **PP.** *Two straight lines that intersect one another cannot be both parallel to the same straight line.*

Or, in another common wording,

PP. *Given a straight line and a point not on the line, there is at most one straight line through the point that does not intersect the given line.*

Playfair's Parallel Postulate got its current name from the Scottish mathematician John Playfair (1748–1819), who brought out successful editions of Euclid's *Elements* in the years following 1795. After 1800 many commentators referred to Playfair's postulate (**PP**) as the best statement of Euclid's postulate, so it became a tradition in many geometry books to use **PP** instead of **EFP**. The reader can see easily that both **PP** and **EFP** assert the uniqueness of parallel lines and neither asserts the existence of parallel lines. **HSP** seems to have been first stated by Hilbert in the first edition of his *Grundlagen der Geometrie* (but not in the later editions, where he used **PP**) and in [**EG**: Hilbert]. But later, apparently starting in the 1960s, many (but not all) textbooks (including earlier editions of this book!) started calling **HSP** by the name "Playfair's Postulate". In absolute geometry **HSP** is equivalent to **PP** and **EFP** (as you showed in **10.3a**), and because absolute geometry was the focus of many investigations, the statement "**HSP** is equivalent to **EFP**" was made and is still being repeated in many textbooks and expository writings about geometry even when the context is not absolute geometry. As you showed in **10.4a**, **EFP** is true on spheres and so is **PP**, and so they cannot be equivalent to **HSP**, which is clearly false on the sphere.

During the 9th through 12th centuries, parallel postulates were explored by mathematicians in the Islamic world. al-Hasan ibn al-Haytham (Persian, 965–1040) proved **EFP** by assuming, *A quadrilateral with three right angles must have all right angles.* Quadrilaterals with three right angles are known later in Western literature as Lambert quadrilaterals after Johann Lambert (German–Swiss, 1728–1777), who studied them and also was the first to extensively investigate hyperbolic trigonometric functions. The Persian poet and geometer Omar Khayyam (1048–1131) wrote a book that in translation is entitled *Discussion of Difficulties in Euclid* [**AT**: Khayyam 1958]. Khayyam introduces a new postulate (which he attributes to Aristotle, though it has not been found among the surviving works of Aristotle), which says *Two straight lines that start to converge continue to converge.* In his work on parallel lines he studied the Khayyam quadrilaterals (later in the West to be called Saccheri quadrilaterals), which are discussed here in Chapter 12. Nassir al-Din Al-Tusi (Khorasan, 1201–1274) furthered the study of parallel postulates and is credited (though some say it may have been his son) with first proving that $A = 180$ *is equivalent to* **EFP**, our **10.3bc**. Al-Tusi's works were the first Islamic mathematical works to be discovered in the Western Renaissance and were published in Rome in 1594.

The assumption *parallel lines are equidistant* (see our **10.2d**) was discussed in various forms by Aristotle (Greek, 383–322 B.C.), Posidonius (Greek, 135–51 B.C.), Proclus, ibn Sina (Uzbek, 980–1037), Omar Khayyam, and Saccheri (Italian, 1667–1733).

John Wallis (English, 1616–1703) proved that **EFP** followed from the assumption *To every triangle, there exists a similar triangle of arbitrary magnitude.* (See our **9.4**.) The works, already mentioned, of Wallis, Saccheri, Lambert in the 17th and 18th centuries were continued by the French school into the early 19th century by Joseph Fourier (French, 1768–1830), who concluded that geometry was a physical science and could not be established a priori, and Adrien-Marie Legendre (1752–1833), who proved that *in absolute geometry the sum of the interior angles of a triangle is always less than or equal to 180°.*

The breakthrough in the study of parallel postulates came in the 19th century when, apparently independently, János Bolyai (Hungarian, 1802–1860), and N. I. Lobachevsky (Russian, 1792–1856) finally developed hyperbolic geometry as an absolute geometry that did not satisfy **EFP**. See Chapters 5 and 17 for more discussions of hyperbolic geometry. These discoveries of hyperbolic geometry showed that the quest for a proof of **EFP** from within absolute geometry is impossible.

For details of the relevant histories, see [**HI**: Gray], [**DG**: McCleary], [**HI**: Rosenfeld], and the Heath's editorial notes in [**AT**: Euclid, *Elements*].

NON-EUCLIDEAN GEOMETRIES WITHIN THE HISTORICAL STRANDS

Building Structures Strand. Because of this long history of investigation into parallel postulates within the Building Structures Strand, many books misleadingly call hyperbolic geometry "*the* non-Euclidean geometry." As pointed out in the introduction to Chapter 2, spherical geometry has been studied since ancient times within the Navigation/Stargazing Strand, but it did not fit into absolute geometry and thus was (and still is) left out of many discussions of non-Euclidean geometry, especially those that take place in the context of the Building Structures Strand. In those discussions that do include spherical geometry, hyperbolic geometry is usually called *Lobachevskian geometry* or *Lobachevskian/Bolyai geometry*, and spherical geometry is usually called *Riemannian geometry*. Bernhard Riemann (German, 1826–1866) pioneered an intrinsic and analytic view for surfaces and space and suggested that the intrinsic geometry of the sphere could be thought of as a non-Euclidean geometry. He also introduced an intrinsic analytic view of the sphere that became known as the *Riemann sphere*. The Riemann sphere is usually studied in a course on complex analysis. The use of the term "Riemannian geometry" to denote the non-Euclidean geometry of the sphere leads to confusions since "Riemannian geometry" more commonly refers to the part of differential geometry that gives intrinsic analytic descriptions of the local geometry of general surfaces and 3-space.

Art/Pattern Strand.   In this strand, artists' experimentations with perspective in art led to the mathematical theory of *projective geometry* that developed initially independently of both spherical and hyperbolic geometries. Then in the last part of the 19th century, projective geometry was used to provide a unified treatment of the three geometries. It was within this development that Felix Klein introduced the names *double elliptic geometry* (for the geometry of the sphere), *elliptic geometry* (for the geometry of the sphere with antipodal points identified — this the same as the real projective plane, **RP²**, discussed in Problem **18.3**), *parabolic geometry* (for Euclidean geometry), and *hyperbolic geometry* (for the Bolyai/Lobachevsky geometry). In Chapter 11, we will discuss Klein's development of various geometries in terms of their isometries. A geometric connection between spherical and projective geometry can be seen in the image onto the plane of the sphere (with antipodal points identified) under gnomic projection. (See Problems **14.2** and **20.6**.)

For a fuller discussion of spherical and hyperbolic geometry in this strand, see [**HI**: Kline], Chapter 38, and [**HI**: Katz], Section 17.3.

In Kline we also find a statement (page 913) that represented the view of some mathematicians:

> By the early 1870s several basic non-Euclidean geometries, the hyperbolic, and the two elliptic geometries, had been introduced and intensively studied. The fundamental question which had yet to be answered in order to make these geometries legitimate branches of mathematics was whether they were consistent.

So for those who believe that the only "legitimate" branches of mathematics are those that have been included in some rigorous formal system, then non-Euclidean geometries did not became a part of mathematics until the late 1800s.

Navigation/Stargazing Strand. In the introduction to Chapter 2 we saw that spherical geometry has been studied within this strand for more than 2000 years, and in this study were developed (spherical) coordinate systems and trigonometric functions. We hope that the reader agrees that this was (and still is) legitimate mathematics.

For more readings (and references) on this history of spherical geometry within the Navigation/Stargazing Strand, see [**HI**: Katz], Chapter 4, and [**HI**: Rosenfeld], Chapter 1. Into the 19th century, spherical geometry occurred almost entirely in the Navigation/Stargazing Strand and was used by Brahe and Kepler in studying the motion of stars and planets and by navigators and surveyors. The popular book (which is also available online) *Spherical Trigonometry: For the use of colleges and schools* (1886) [**SP**: Todhunter] contains several discussions of the use of spherical geometry in surveying and was used in British schools before hyperbolic geometry was widely known in the British Isles. On the continent, C. F. Gauss (1777–1855) was using spherical geometry in various large-scale surveying projects before the advent of hyperbolic geometry. Within this context, Gauss initiated (and Riemann expanded) the analytic study of surfaces that led to what we now call differential geometry. Gauss introduced *Gaussian curvature*, which was discussed in Chapter 7. Within this theory of surfaces, spherical geometry is described locally as the geometry of surfaces with constant positive (Gaussian) curvature and hyperbolic geometry is described locally as the geometry of surfaces with constant negative curvature. As we discussed in Chapter 5, there are no analytic surfaces in 3-space with the complete hyperbolic geometry, but the pseudosphere (Problem **5.3**) locally has hyperbolic geometry. Because of this, in the Navigation/Stargazing strand, hyperbolic geometry is called by some authors ***pseudosphere geometry***.

Chapter 11

ISOMETRIES AND PATTERNS

What geometrician or arithmetician could fail to take pleasure in the symmetries, correspondences and principles of order observed in visible things?
— Plotinus, *The Enneads*, II.9.16 [**AT**: Plotinus]

[Geometry is] the study of the properties of a space which are invariant under a given group of transformations.
F. Klein, *Erlangen Program*

Life forms illogical patterns. It is haphazard and full of beauties which I try to catch as they fly by, for who knows whether any of them will ever return?   — Margot Fonteyn

Recall that in Chapter 1 we gave the following:

DEFINITION. An *isometry* is a transformation that preserves distances and angle measures.

In this chapter we will show (for the plane, spheres, and hyperbolic planes) that every isometry is the composition (product) of (not more than three) reflections, and we will determine all the different types of isometries. This finishes the study of reflections and rotations we started in Problem **5.4**. We will note the differences between the kinds of isometries that appear in the three geometries.

Then we will study patterns in these three spaces. Along the way we will look at some group theory through its origins, that is, geometrically.

This material originated primarily within the Art/Pattern Strand.

It would be good for the reader to start by reviewing Problem **5.4**. We will start the chapter with a further investigation of isometries and then with a discussion of definitions and terminology. We advise the reader to investigate this introductory material as concretely as possible, making drawings and/or moving about paper triangles.

PROBLEM **11.1**   ISOMETRIES

Definitions:
*A **reflection through the line** (**geodesic**) l is an isometry* R_l *such that it fixes* only *those points that lie on l and, for each point P not on l, l is the perpendicular bisector of the geodesic segment joining P to* $R_l(P)$.

Reflections are closely related to our notion of geodesic (instrinsically straight). We say in Chapters 1, 2, 4, and 5 that the existence of (local) reflections through any geodesic is a major determining property of geodesic in the plane, spheres, cylinders, cones, and hyperbolic planes.

*A **directed angle** is an angle with one of its sides designated as the **initial side** (the other side is then the **terminal side**). It is usual to indicate the direction with an arrow, as in Figure 11.1. The usual convention is to say that angles with a counter-clockwise direction are positive and those with a clockwise direction are negative. When we write $\angle APB$ for a directed angle, then we consider AP to be on the initial side*

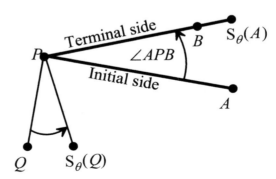

Figure 11.1 Directed angle and rotation about P through angle $\angle APB$

*A **rotation about P through the directed angle** θ is an isome-try* S_θ *that leaves the point P fixed and is such that, for every Q ≠ P, $S_\theta(Q)$ is on the same circle with center P that Q is on, and the angle ∠QPS$_\theta$(Q) is congruent to θ and in the same direction.* See Figure 11.1.

    a. Prove: *If P is a point on the plane, sphere, or hyperbolic plane and ∠APB is any directed angle at P, then there is a rotation about P through the angle θ = ∠APB.*

Refer to Problem **5.4** and Figure 5.13. Remember to show that the composition of these two reflections has the desired angle property. Remember to check that every point is rotated.

    But, it is still not yet clear that anything we would want to call a translation exists on spheres or hyperbolic planes.

    b. *Let m and n be two geodesics on the plane, a sphere, or a hyperbolic plane with a common perpendicular l. Look at the composition of the reflection R_m through m with the reflection R_n through n. Show why this composition R_nR_m could be called a translation of the surface along l. How far are points on l moved? What happens to points not on l?* See Figure 11.2.

Remember that R_nR_m denotes: *First* reflect about *m* and *then* reflect about *n*. Let *Q* be an arbitrary point on *l* (but not on *m* or *n*). Investigate where *Q* is sent by R_m and then by R_nR_m. Then investigate what happens to points not on *l* — note that they stay the same distance from *l*. Be sure to draw pictures.

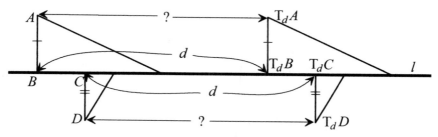

Figure 11.2 Translation of distance *d* along *l*

We can now formulate a definition that works on all surfaces:

*A **translation of distance d along the line** (**geodesic**) l is an isome-try* T*$_d$ that takes each point on l to a point on l at the distance (along l) of d and takes each point not on l to another point on the* same side *of l and at the same distance from l.* See Figure 11.2.

Note that on a sphere a translation along a great circle *l* is the same as a rotation about the poles of that great circle.

    c.   Prove: *If l is a geodesic on the plane, a sphere, or a hyperbolic plane and d is a distance, then there is a translation of distance d along l.*

Use part **b**.

In Figure 11.3 (which can be considered to be on either the plane, a sphere, or a hyperbolic plane), two congruent geometric figures, $\mathscr{F}$ and $\mathscr{G}$, are given, but there is not a single reflection, or rotation, or translation that will take one onto the other.

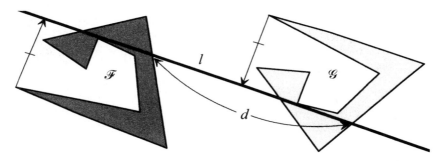

Figure 11.3   Glide reflection

However, it is clear that there is some composition of translations, rotations, and reflections that will take $\mathscr{F}$ onto $\mathscr{G}$. In fact, the composition of a reflection through the line *l* and a translation along *l* will take $\mathscr{F}$ onto $\mathscr{G}$. This isometry is called *a **glide reflection along l**.*

*A **glide reflection** (or just plain **glide**) **of distance d along the line** (**geodesic**) l is an isometry* G*$_d$ that takes each point on l to a point on l at the distance (along l) of d and takes each point not on l to another point on the* other side *of l and at the same distance from l.*

d. *If l is a geodesic on the plane, sphere, or hyperbolic plane and if d is a distance, then there is a glide of distance d along l.*

Later in this chapter we will show that these are the only isometries of the plane and spheres. However, on the hyperbolic plane there is another isometry that is not a reflection, rotation, translation, or glide.

On the hyperbolic plane there are some pairs of geodesics, called **asymptotic geodesics**, that do not intersect and also do not have a common perpendicular (and thus are not parallel transports). See Problem **8.4b**. For example, two radial geodesics in the annular hyperbolic plane are asymptotic. The composition of reflections about two asymptotic geodesics is defined to be a **horolation**. See Figure 11.4, where $A' = R_m(A)$, $A'' = R_n(R_m(A))$, $B'' = R_n(R_m(B))$, $C'' = R_n(R_m(C))$.

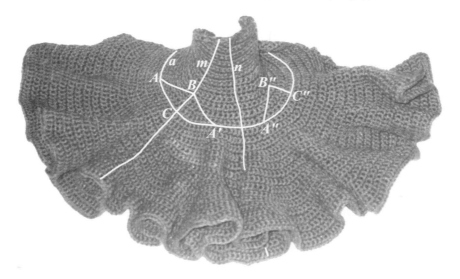

Figure 11.4   Horolation

Note that every point moves along an annular line (not a geodesic) such as *a* in Figure 11.4. The annular lines could be called "circles of infinite radius". [*Do you see why? If you are given a circular arc in the plane how to you find its center?*] In much of the literature these annular lines, or circles of infinite radius, are called **horocycles**. [In the past, some students have affectionately call them "horrorcycles."] Note that *on the plane circles of infinite radius are straight lines.*

For the horolation depicted in Figure 11.4 the two reflection lines are radial geodesics. This is not really the special case it looks to be: If l and m are any two asymptotic (but not radial) geodesics, then l must intersect a radial geodesic r, in fact, infinitely many radial geodesics. Reflect the whole hyperbolic plane through the bisector b of the angle between the end of l at which it is asymptotic to m and the end of r at which it is asymptotic to other radial geodesics. See Figure 11.5. The images of l and m under the reflection are now radial geodesics, $r = R_b(l)$ and $R_b(m)$.

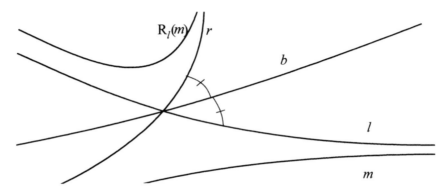

Figure 11.5   Asymptotic geodesics can be reflected to radial geodesics

A horolation is an isometry that is neither a rotation nor a translation but can be thought of as a rotation about the point at infinity where the two asymptotic lines converge. In the case of radial geodesics, a horolation will take each annulus to itself because the radial geodesics are perpendicular to the annuli. See Figure 11.4.

In Chapter 17 we will also see that a horolation corresponds to a translation parallel to the x-axis in the hyperbolic coordinate system introduced in Problem **5.2** and the upper half-plane model introduced in Problems **17.1** and **17.2**. In the context of the models of the hyperbolic plane introduced in Chapter 17 many authors use the terms "elliptic isometry", "parabolic isometry", and "hyperbolic isometry" to refer to what we are calling, "rotation", "horolation", and "translation."

PROBLEM 11.2 THREE POINTS DETERMINE AN ISOMETRY

In order to analyze what all isometries are, we need the following very important property of isometries:

Prove the following: On the plane, spheres, or hyperbolic planes, if *f* and *g* are isometries and *A*, *B*, *C* are three non-collinear points, such that *f*(*A*) = *g*(*A*), *f*(*B*) = *g*(*B*), and *f*(*C*) = *g*(*C*), then *f* and *g* are the same isometry, that is, *f*(*X*) = *g*(*X*) for every point *X*.

Let us see an example of how this works before you see why this property holds. In Figure 11.6, H_P represents the half-turn around *P*, R_m represents the reflection through line *m*, G_l represents a glide reflection along *l*, and

$$\mathscr{F}_1 = \{A, B, C\}, \quad \mathscr{F}_2 = \{A', B', C'\} \quad \text{and} \quad \mathscr{F}_3 = \{A'', B'', C''\}.$$

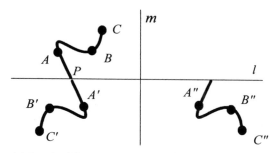

Figure 11.6   A glide equals a half-turn followed by a reflection

We can see that $H_P(\mathscr{F}_1) = \mathscr{F}_2$, $R_m(\mathscr{F}_2) = \mathscr{F}_3$, and $G_l(\mathscr{F}_1) = \mathscr{F}_3$. But then G_l and $R_m H_p$ perform the same action on the three points. If we apply the above result, we can say that $G_l = R_m H_P$; that is, $G_l(X) = R_m H_P(X)$, for all points *X* on the plane. You can see now the usefulness of proving Problem **11.2**.

If you have trouble getting started with this problem, then take a specific example of two congruent triangles △*ABC* and △*A'B'C'* such as in Figure 11.7. Pick another point *X* and convince yourself as concretely as possible that there is only one location that *X* will be taken to by any isometry that takes *A*, *B*, *C* onto *A'*, *B'*, *C'*.

PROBLEM **11.3.** CLASSIFICATION OF ISOMETRIES

a. Prove that on the plane, spheres, or hyperbolic planes, every isometry is the composition of one, two, or three reflections.

Look back at what we did in the proofs of SAS and ASA. Also use Problem **11.2**, which allows you to be more concrete when thinking about isometries because you only need to look at the effect of the isometries on any three non-collinear points that you pick. To get started on **11.3**, cut a triangle out of an index card and use it to draw two congruent triangles in different orientations on a sheet of paper. For example, see Figure 11.7. Can you move one triangle to the other by three (or fewer) reflections? You can use your cutout triangle for the intermediate steps.

Figure 11.7   Can you make triangles coincide with three reflections?

We have already showed in Problems **5.4** and **11.1** that the product of two reflections is a rotation (if the two reflection lines intersect) and a translation (if the two reflection lines have a common perpendicular, that is, if the two lines are parallel transports). Thus we can

> **b.** *Prove that on the plane and spheres, every composition of two reflections is either the identity, a translation, or a rotation. What are the other possibilities on a hyperbolic plane? Why are you sure that these are the only isometries that are the composition of two reflections? What happens if you switch the order of the two reflections?*

In this part start with the two reflections as given. What are the different ways in which the two reflection lines can intersect? To fully answer this part on a hyperbolic plane, you will have to use the result (found informally in Chapters 5 and 8 and proved in Problem **17.3**) that *Two geodesics in a hyperbolic plane either intersect in a point, or have a common perpendicular, or are asymptotic.*

Your proof of part **b** or Problem **11.1** probably already shows that

THEOREM 11.3. *On the plane, spheres, and hyperbolic planes, a rotation is determined by two intersecting reflection lines (geodesics). The lines are determined only by their point of*

intersection and the angle between them; that is, any two lines with the same intersection point and the same angle between them will produce the same rotation. See Figure 11.8.

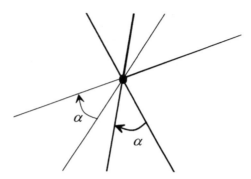

Figure 11.8   Two pairs of reflection lines, which determine the same rotation

   c. *On the plane, spheres, and hyperbolic plane, the product of two rotations* (in general, about different points) *is a single rotation, translation, or horolation. Show how to determine geometrically which specific isometry you obtain, including the center and angle of any rotation.*

Use Theorem 11.3 and write the composition of the two rotations with each rotation the composition of two reflections in a way that the middle two reflections cancel out. Be careful to keep track of the directions of the rotations. See Figure 11.9.

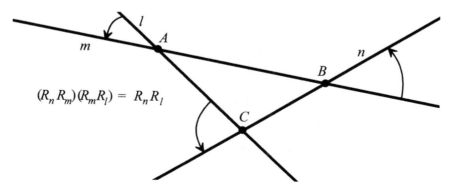

$$(R_n R_m)(R_m R_l) = R_n R_l$$

Figure 11.9 Composition of two rotations

d. *On the plane, spheres, and hyperbolic planes, every composition of three reflections is either a reflection or a glide reflection. How can you tell which one?*

Approach using Theorem 11.3: Write the composition of three reflections, R_l, R_m, R_n, as either $(R_lR_m)R_n$ or $R_l(R_mR_n)$. If the reflection lines, l, m, n, intersect, then use part **c** to replace the reflection lines within the parentheses with reflection lines that produce the same result. Try to produce the situation where the first (or last) reflection line is perpendicular to the other two. What is the situation if none of l, m, n intersect each other? This approach works well on the plane and spheres but is more difficult on a hyperbolic plane because the situation when all three lines do not intersect is more complicated.

Approach using Problem 11.2 and triangles: Let $\triangle ABC$ be a triangle and let $\triangle A'B'C'$ be its image under the composition of the three reflections. Note that the two triangles cannot be directly congruent (see the discussion around Figure 6.5). Extend two corresponding sides (say AB and $A'B'$) of the triangles to lines (geodesics) l and m. Then there are three cases: The lines l and m intersect, or have a common perpendicular, or are asymptotic.

If parts **a**, **b**, and **d** are true, then

e. *Every isometry of the plane or a sphere or a hyperbolic plane is either a reflection, a translation, a rotation, a horolation, a glide reflection, or the identity.*

Notice that in our proofs of SAS and ASA and (probably) in your proof of part **b** we only need to use reflections about the perpendicular bisectors or segments joining two points. Because, on a sphere, we only need great circles to join two points and for perpendicular bisectors, we can restate part **b** on a sphere to read

Every isometry of a sphere is the composition of one, two, or three reflections through great circles.

Thus, in particular, if there were a reflection of the sphere that was not through a great circle, then that reflection (being an isometry) would also be composition of one, two, or three reflections through great circles.

You now have powerful tools to make a classification of discrete strip patterns on the plane, spheres, and hyperbolic planes and finite patterns on the plane and hyperbolic planes.

KLEIN'S ERLANGEN PROGRAM

In 1872, on the occasion of his appointment as professor in the University of Erlangen at the age of 23, Felix Klein (1849–1925) proposed that the view of symmetries should be radically extended. He proposed viewing geometry as "the study of the properties of a space which are invariant under a given group of transformations." The proposal is now universally known as the *Erlangen Program*.

Klein provided several examples of geometries and their associated transformations. In Euclidean plane geometry the transformations are all the isometries (reflections, rotations, translations, and glides) and including the similarity transformations (dilations), that are not isometries because they do not preserve distances but do preserve angles and thus take triangles to similar triangles (see Problem **13.4**).

Within the Erlangen Program, spherical geometry is the study of the isometries of the sphere (reflections, rotation/translations, and glides) and hyperbolic geometry is the study of its isometries (reflections, rotations, translations, glides, and horolations). In general, symmetry was used as an underlying principle and it was shown that different geometries could coexist because they dealt with different types of propositions and invariances related to different types of (symmetry) transformations. Topology, projective geometry, and other theories have all been placed within the classification of geometries in the Erlangen Program. The long-term effects of the Erlangen Program can be seen all over pure mathematics, and the idea of transformations and of synthesis using groups of symmetry is now standard also in physics. Klein's demonstration of the relationship of geometry to groups of transformations helped to provide impetus for the development of the abstract notion of a *group* by the end of 19th century. For more discussion of Klein's program, see [**HM**: Yaglom].

SYMMETRIES AND PATTERNS

In Chapter 1 we talked about symmetries of the line. All of those symmetries can be seen as isometries of the plane except for similarity symmetry and 3-D rotation symmetry (through any angle not an integer multiple of 180°). Similarity symmetry changes lengths between points of the geometric figure and thus is not an isometry. Three-dimensional rotation symmetry is an isometry of 3-space, but it moves any plane off itself and thus cannot be an isometry of a plane (unless the angle of rotation is a multiple of 180°). The notion of symmetry grew out of the Art/Pattern Strand of history. Well before written history, symmetry and patterns were part of the human experience of weaving and decorating.

What is a symmetry of a geometric figure? A *symmetry* of a geometric figure is an isometry that takes the figure onto itself. For example, reflection through any median is a symmetry of an equilateral triangle. See Figure 11.10.

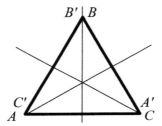

Figure 11.10   Reflection symmetries

It is easy to see that rotations through 1/3 and 2/3 of a revolution, $S_{1/3}$ and $S_{2/3}$, are also symmetries of the equilateral triangle. In addition, the identity, Id, is (trivially, by the definition) an isometry and, thus, is also a symmetry of the equilateral triangle. Therefore, the equilateral triangle has six symmetries: R_A, R_B, R_C, $S_{1/3}$, $S_{2/3}$, Id, where R_A, R_B, R_C, denote the reflections through the medians from A, B, C. These are the only symmetries of the equilateral triangle.

Now look at the geometric figure in Figure 11.11. It has exactly the same symmetries as an equilateral triangle. Though the two figures look very different, we say they are *isomorphic patterns*.

A *pattern* is a figure *together with all its symmetries*; and we call the collection of all symmetries of a geometric figure its *symmetry group*. We want some way to denote that two different patterns have the same symmetries, as is the case with the pattern in Figure 11.11 and the equilateral triangle in Figure 11.10. We do this by saying that two patterns are *isomorphic* if they have the same symmetries. This should become clearer through more examples.

Figure 11.11   Same symmetries as the equilateral triangle

The letters S and N each have only half-turn and the identity as symmetries; thus we say they are isomorphic patterns with symmetry group $\{Id, S_{1/2}\}$. Similarly, the letters A and M each have only vertical reflection and the identity as symmetries and thus are isomorphic patterns with symmetry group $\{Id, R\}$. Note that the letters S and A each have the same number of symmetries, but we do **not** call them isomorphic patterns because the symmetries are different symmetries. The reader should check their understanding by finding isomorphic patterns among the other letters of the alphabet in normal printing by hand.

To construct further examples, we can start with a geometric figure, often called a *motif*, that has no symmetry (except the identity). For example, the geometric figures in Figure 11.12 are possible motifs.

Figure 11.12   Motifs

To make examples of patterns with a specific isometry, we can start with any motif and then use the isometry and its inverse to make additional copies of the motif over and over again. In the process we obtain another geometric figure for which the initial isometry is a symmetry. Let us look at an example: If we start with the first motif in Figure 11.12 and the isometry is translation to the right through a distance d, then, using this isometry and its inverse (translation to the left through a distance d) and repeating them over and over, we obtain the pattern in Figure 11.13.

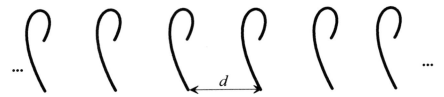

Figure 11.13   Pattern with translation symmetries

The symmetry group of the pattern in Figure 11.13 is

$$\{\text{Id (the identity)}, \text{T}_{nd} \text{ (where } n = \pm 1, \pm 2, \pm 3, ...)\}.$$

If the isometry is clockwise rotation through 1/3 of a revolution about the lower endpoint of the motif, then we obtain the pattern in Figure 11.14. This figure is a pattern with symmetries $\{\text{Id}, \text{S}_{1/3}, \text{S}_{2/3}\}$. Note that this pattern is not isomorphic with the equilateral triangle pattern.

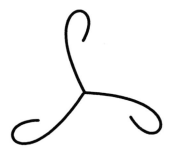

Figure 11.14   Rotation symmetry

If, in the constructions depicted in Figures 11.13 and 11.14, we replace the motif with any other motif (with no non-trivial symmetries), then we will get other patterns that are isomorphic to the original ones because the symmetries are the same.

We call the collection of all symmetries of a geometric figure its *symmetry group*. If g, h are symmetries of a figure $\mathscr{F}$, then you can easily see that

*The **composition** gh (first transform by h and then follow it by g) is also a symmetry.* For example, for an equilateral triangle the composition of the reflection R_A with the reflection R_B is a rotation, $S_{2/3}$. In symbols, $R_B R_A = S_{2/3}$. See Figure 11.15.

*Composition of symmetries is **associative**: That is, if h, g, k are symmetries of the same figure, then (hg)k = h(gk) ≡ hgk.*

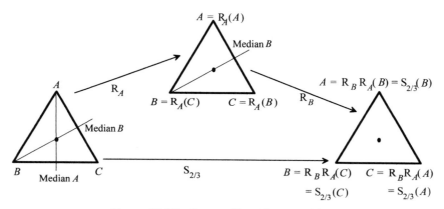

Figure 11.15   Composition of symmetries

*The **identity** transformation, Id (the transformation that takes every point to itself), is a symmetry.*

*For every symmetry g of a figure there is another symmetry f such that gf and fg are the identity — in this case we call f the **inverse** of g.* For example, the inverse of the rotation $S_{2/3}$ is the rotation $S_{1/3}$ and vice versa. In symbols, $S_{1/3} S_{2/3} = S_{2/3} S_{1/3} = $ Id.

Those readers who are familiar with abstract group theory will recognize the above as the axioms for an abstract group. We will discuss further connections with group theory at the end of this chapter. If two patterns are isomorphic, then their symmetry groups are isomorphic as abstract groups. The converse is often, but not always, true. For example, the symmetry groups of the letter A and the letter S are isomorphic (as abstract groups) to Z_2, but they are not isomorphic as patterns because they have different symmetries.

A **strip** (or linear, or frieze) **pattern** is a pattern that has a translation symmetry, with all of its symmetries also symmetries of a given line. For example, the pattern in Figure 11.13 is a strip pattern as is the pattern in Figure 11.11. The strip pattern in Figure 11.16 has symmetry group: $\{Id, R_l, T_{nd}, G_{nd} = R_l T_{nd}$ (where $n = 0, \pm1, \pm2, \pm3, \cdots)\}$.

Figure 11.16   A strip pattern

You are now able to start to study properties about patterns and isometries.

PROBLEM 11.4   EXAMPLES OF PATTERNS

a. *Go through all the letters of the alphabet* (in normal printing by hand) *and decide which are isomorphic as patterns.*

b. *Find as many* (non-isomorphic) *patterns as you can that have only finitely many symmetries. List all the symmetries of each pattern you find.*

c. *Find as many* (non-isomorphic) *strip patterns as you can. List all the symmetries of each strip pattern you find.*

The purpose of this problem is to get you looking at and thinking about patterns. Examples of different strip patterns and many finite patterns can be found on buildings everywhere: houses of worship, courthouses, and most older buildings. Also look for other decorations on plates or on wallpaper edging. In the next two problems we will determine if your list of strip patterns and finite patterns contains all possible strip and finite patterns.

*PROBLEM 11.5   CLASSIFICATION OF DISCRETE STRIP PATTERNS

A strip pattern is **discrete** if every translation symmetry of the strip pattern is a multiple of some shortest translation.

a. *Prove there are only seven non-isomorphic strip patterns on the plane that are discrete.*

b. *What are some non-discrete strip patterns?*

c. *What happens with strip patterns on spheres and hyperbolic planes?*

Hint: Use Problem **11.3**.

*PROBLEM **11.6** CLASSIFICATION OF FINITE PLANE PATTERNS

Look at all the finite patterns on the plane that you found in Problem **11.5b**. Do you notice that for each there is a point (not necessary on the figure) such that every symmetry of the pattern leaves the point fixed? See Figure 11.17.

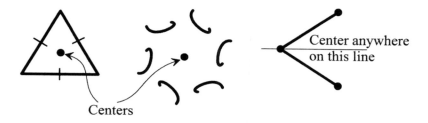

Figure 11.17   Finite patterns have centers

This leads to the following:

a. *Show that any pattern on the plane with only finitely many symmetries has a center. That is, there is a point in the plane (not necessarily on the figure) such that every symmetry of the pattern leaves the point fixed. Is this true on spheres and hyperbolic planes?*

This was first proved by Leonardo da Vinci (1452–1519, Italian), and you can prove it too! *Hint:* Start by looking at what happens if there is a translation or glide symmetry.

b. *Describe all the patterns on the plane and hyperbolic planes with only finitely many symmetries.*

Hint: Use part **a**. What rotations are possible if there are only finitely many symmetries?

c. *Describe all the patterns on the sphere that are finite and that have centers.*

Hint: If there is one center, then its antipodal point is necessarily also a center. *Why?*

If you take a cube with its vertices on a sphere and project from the center of the sphere the edges of the cube onto the sphere, then the result is a pattern on the sphere with only finitely many symmetries. This pattern does not fit with the patterns you found in part **c** because this pattern has no center (on the sphere). See Problems **11.7b** and **23.5** for more examples.

*PROBLEM 11.7   REGULAR TILINGS WITH POLYGONS

We will now consider some special infinite patterns of the plane, spheres, and hyperbolic planes. These special patterns are called *regular tilings* (or *regular tessellations*, or *mosaics*).

> **DEFINITION.** *A regular tiling {n, k} of a geometric space (the plane, a sphere, or a hyperbolic plane) is made by taking identical copies of a regular n-gon (a polygon with n edges) and using these n-gons to cover every point in the space so that there are no overlaps except each edge of one n-gon is also the edge of another n-gon and each vertex is a vertex of k n-gons.*

You probably know the three familiar regular tilings of the plane: {3, 6}, the regular tiling of the plane by triangles with six triangles coming together at a vertex; {4, 4}, the regular tiling of the plane by squares with four squares coming together at a vertex; and {6, 3}, the regular tiling of the plane by hexagons with three hexagons coming together at each vertex. There is only one way to tile a plane with regular hexagons; however, there are other ways to tile a plane with regular triangles and squares but only one for each that is a *regular* tiling.

Note that each regular tiling can also be thought of as a (infinite) pattern. The notation $\{n, k\}$ is called the **Schläfli symbol** of the tiling, named after Ludwig Schläfli (1814–1895, Swiss). We now study the possible regular tilings.

    a. *Show that, $\{3, 6\}$, $\{4, 4\}$, $\{6, 3\}$ are the only Schläfli symbols that represent regular tilings of the plane.*

Focus on what happens at the vertices.

    b. *Find all the regular tilings of a sphere.*

Again, focus on what happens at the vertices. Remember that angles of a regular n-gon on the sphere are larger than the angles of the corresponding regular n-gon on the plane (*Why?*); and use Problems **7.1** and **7.4**.

There are more finite patterns of the sphere besides those given in Problem **11.6c** and **11.7b**. See a soccer ball for an example and [**SG:** Montesinos] for the complete classification.

    c. *Show that each of the Schläfli symbols $\{n, k\}$ with both n and k greater than 1 represents a regular tiling of the plane* (part **a**), *or of a sphere* (part **b**), *or of a hyperbolic plane.*

Again, focus on what happens at the vertices.

*OTHER PERIODIC (AND NON-PERIODIC) PATTERNS

There are numerous non-regular tilings and other patterns other than regular tilings. Besides the three that come from regular tilings, there are 14 more (non-isomorphic) infinite periodic patterns on the plane. ("Periodic" means that the pattern has a minimal translation symmetry.) See Figure 11.18. These 17 periodic patterns in the plane are often called **wall-paper patterns**. See Escher's (Maurits Cornelius Escher, 1898–1972, Dutch) drawings for examples and [**SG:** Budden] for proofs and more examples. For a complete exposition on periodic patterns and tilings on the plane, see [**SG:** Grünbaum].

Figure 11.18 A repeating pattern that is not a regular tiling
— the thin straight lines (forming squares) are lines of reflection symmetry,
there is half-turn symmetry about corners of the squares, and
there is 4-fold rotation symmetry about the center of each square.

There also exist non-periodic tilings of the plane; see Figure 11.19. Notice the 5-fold local rotation symmetry at the center — this will never repeat anywhere in the pattern and thus the pattern cannot be periodic.

Figure 11.19 Non-periodic tiling of the plane

For more discussion of non-periodic tilings, see [**SG**: Senechal] and [**SG**: Grünbaum].

See [**SG**: Montesinos] for a discussion about other patterns and tilings on hyperbolic spaces.

There are 230 non-isomorphic repeating patterns in Euclidean 3-space. These are the patterns that appear in crystals. However, in protein crystals there are no reflection (through planes) or central (reflection through a point) symmetries — as a consequence there are only 65 repeating patterns without plane reflections or central symmetries in 3-space. The corresponding groups are call ***chiral groups***. See [SG: Giacovazzo] for more information on crystallographic groups.

*GEOMETRIC MEANING
OF ABSTRACT GROUP TERMINOLOGY

The collection of symmetries of a geometric figure with the operation of composition is an ***abstract group.*** We showed above how the usual axioms of a group are satisfied. For example, the pattern in Figure 11.20 has symmetry group $\{R_A, R_B, R_C, S_{1/3}, S_{2/3}, Id\}$ that is isomorphic as an abstract group to what is usually called $\mathbf{D_3}$, the third dihedral group.

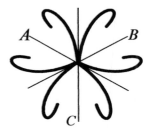

Figure 11.20   Isomorphic to **D₃**

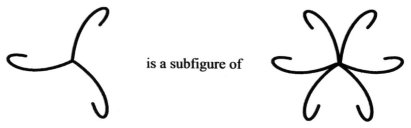

is a subfigure of

Figure 11.21   Subgroups

If a figure has a subfigure and if the symmetries of the subfigure are all also symmetries of the original figure, then the symmetry group of the subfigure is a ***subgroup*** of the symmetry group of the original figure. See Figure 11.21.

The symmetry group of the subfigure is {$S_{1/3}$, $S_{2/3}$, Id}, which is isomorphic as an abstract group to Z_3. This group is a subgroup of the symmetry group of the original figure, {$S_{1/3}$, $S_{2/3}$, R_A, R_B, R_C, Id}, which is isomorphic to D_3.

In D_3 the two *cosets* of Z_3, Id Z_3 and $R_AZ_3 = R_BZ_3 = R_CZ_3$, correspond to the two copies of the subfigure in Figure 11.22.

Figure 11.22   Two cosets of the subgroup in the group

In general, if G is the symmetry group of a figure and H is a subgroup that is the symmetry group of a subfigure in the figure, then the *cosets* of H correspond to the several congruent copies of the subfigure that exist within the larger figure.

The term *group* was coined (as *groupe* in French) by E. Galois (1811–1832) in 1830. The modern definition of a group is somewhat different from that of Galois, for whom the term denoted a subgroup of the group of permutations of the roots of a given polynomial. F. Klein and S. Lie (1842–1899) used the term *closed system* in their earliest writing on the subject of groups. *Group* appears in English in an article by Arthur Cayley, who wrote "A set of symbols, 1, ?, ?, ..., all of them different, and such that the product of any two of them (no matter what order), or the product of any one of them into itself, belongs to the set, is said to be a group." See [**SG**: Lyndon] for more discussion on groups.

Chapter 12

DISSECTION THEORY

Oh, come with old Khayyám, and leave the Wise
To talk, one thing is certain, that Life flies;
One thing is certain, and the Rest is Lies;
The flower that once has blown for ever dies.

— Omar Khayyam, *Rubaiyat*
From the translation by Edward Fitzgerald

WHAT IS DISSECTION THEORY?

Figure 12.1   Parallelogram

In showing that the parallelogram in Figure 12.1 has the same area as a rectangle with the same base and height (altitude), we can easily cut the parallelogram into two pieces and rearrange them to form the rectangle in Figure 12.2.

Figure 12.2   Equivalent by dissection to a rectangle

We say that two figures (F and G) are ***equivalent by dissection*** ($F =_d G$) if one can be cut up into a finite number of pieces and the pieces rearranged to form the other. Some authors use the term "equidecomposable" instead of "equivalent by dissection".

QUESTION: If two planar polygons have the same area, are they equivalent by dissection?

ANSWER: Yes! For all (finite) polygons on either the plane, or on a sphere, or on a hyperbolic plane.

You will prove these results about dissections in this chapter and the next and use them to look at the meaning of area. In this chapter you will show how to dissect any triangle or parallelogram into a rectangle with the same base. Then you will do analogous dissections on spheres and hyperbolic planes after first defining an appropriate analog of parallelograms and rectangles. After that you will show that two polygons on a sphere or on a hyperbolic plane that have the same area are equivalent by dissection to each other. The analogous result on the plane must wait until the next chapter.

The proofs and solutions to all the problems can be done using "$=_d$", but if you wish you can use the weaker notion of "$=_s$": We say that two figures (F and G) are ***equivalent by subtraction*** ($F =_s G$) if there are two other figures, S and S', such that $S =_d S'$ and $F \cup S =_d G \cup S'$, where F ans S and G and S' intersect at most in their boundaries. Some authors use the term "of equal content" instead of "equivalent by subtraction". Saying two figures are equivalent by subtraction means that they can be arrived at by removing equivalent parts from two initially equivalent figures, as in Figure 12.3.

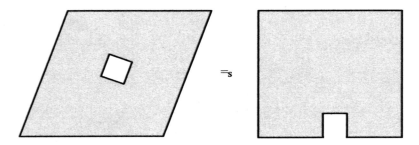

Figure 12.3   Equivalent by subtraction

If we cut out the two small squares as shown in Figure 12.3, we can see that the shaded portions of the rectangle and the parallelogram are equivalent by subtraction, but it is not at all obvious that one can be cut up and rearranged to form the other.

Equivalence by dissection is generally preferable to equivalence by subtraction because it is gives a direct way of seeing that two figures have the same area. However, sometimes it is easier to find a proof of equivalence by subtraction. In addition, equivalence by subtraction has the advantage (as we will see) that in some situations equivalence by dissection is only true if one assumes the Archimedean Postulate (which we first met in Problem **10.3**), while equivalence by subtraction does not need the Archimedean Postulate. However, we would urge you to prove equivalence by dissection wherever you can.

The ***Archimedean Postulate*** (in some books this is called the *Axiom of Continuity*), named after the Greek mathematician Archimedes (who lived in Sicily, 287?–212 B.C.), is as follows:

> ***AP**: *On a line, if the segment AB is less than* (contained in) *the segment AC, then there is a finite* (positive) *integer, n, such that if we put n copies of AB end to end* (see Figure 10.5), *then the nth copy will contain the point C.*

The Archimedean Postulate can also be interpreted to rule out the existence of infinitesimal lengths. It is true that **AP** is needed to prove some results about equivalence by dissection; however, most people assume **AP** to be true on the plane, spheres, and hyperbolic planes.

A DISSECTION PUZZLE FROM 250 B.C. SOLVED IN 2003

About 250 B.C., Archimedes wrote a treatise entitled *Stomachion*, which was lost, though from commentaries it was clear that the work discussed the puzzle pictured in Figure 12.4. This is a puzzle, similar to *Tangrams*, that consists of 14 pieces that fit into a square (all of the vertices lie on a 12 × 12 grid). However, Archimedes' interest in the puzzle was not known. Then in 2003, Reviel Netz, a Stanford historian of science, deciphered parts of the *Palimpsest*, which consists of pieces of parchment that originally (about 1000 A.D.) contained several works of Archimedes, but in the 13th century the words of Archimedes were scraped off and the parchment used to write a prayer book. Netz uncovered the introduction to the *Stomachion* treatise and discovered that Archimedes asked the question *How many different ways can these 14 pieces be rearranged to fit exactly in the square?* Then, on November 17, 2003, it was announced in the *MAA OnLine* column *Math Games*

http://www.maa.org/editorial/mathgames/mathgames_11_17_03.html

that Bill Cutler, a puzzle designer with a Ph.D. in mathematics from Cornell University, found using a computer that there were 17,152 distinct solutions (or only 536 if you counted as the same solutions that varied by rotation or reflection of the square or differed only in the interchange of the congruent pairs of pieces 7 and 14 or 6 and 13). Then, the story was continued on the front page of the *New York Times*, December 14, 2003, where it was announced that the University of California–San Diego mathematicians Ronald Graham and Fan Chung had independently solved the problem using combinatorics. See *MAA OnLine* and the *New York Times* article for more details on the fascinating history of Archimedes' *Palimpsest* and the *Stomachion*.

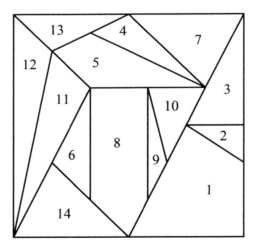

Figure 12.4   Archimedes' Stomachion puzzle

HISTORY OF DISSECTIONS IN THE THEORY OF AREA

Dissections have been the basis, through history, of many proofs for the Pythagorean Theorem, including in Ancient India and China; see also Chapter 13 and Problem **13.2**. In the Plato's *Meno* [AT: Plato] there is a

Socratic dialogue in which is described the dissection proof that the diagonal of a square is the side of a square of twice the area. One of the wall mosaics in the famous Alhambra palace is based on dissection of the square to prove the Pythagorean Theorem. Euclid in his *Elements* implicitly used equivalence by dissection and equivalence by subtraction when he proved propositions about the area (he used the term "equal") of polygons.

The use of dissection to find areas continued after Euclid, but it was not until 1902 that David Hilbert [**FO**: Hilbert] developed Euclid's Postulates and dissection theory into a rigorous theory of area. Hilbert discusses (in his Chapter IV) both equivalence by dissection (*zerlegungsgleich*) and equivalence by subtraction (*ergänzungsgleich*) and when the Archimedean Postulate was necessary. In a footnote, Hilbert gives credit for similar discussions of the theory of areas to M. Gérard (in 1895–1898), F. Schur (in 1898), and O. Stolz (in 1894). A recent detailed discussion of the dissection theory of area can be found in [**TX**: Hartshorne]. A recent history of dissections is contained in Chapter 13.

PROBLEM 12.1   DISSECT PLANE TRIANGLE
AND PARALLELOGRAM

Some of the dissection problems ahead are very simple, while some are rather difficult. If you think that a particular problem was so easy to solve that you may have missed something, chances are you hit the nail right on the head. Most of the dissection proofs will consist of two parts: First show where to make the necessary cuts, and then prove that your construction works, that is, that all the pieces do in fact fit together as you say they do.

a. *Show that on the plane every triangle is equivalent by dissection to a parallelogram with the same base no matter which base of the triangle you pick.*

Part **a** is fairly straightforward, so don't try anything complicated. You only have to prove it for the plane — a proof for spheres and hyperbolic planes will come in a later problem after we find out what to use in place of parallelograms. Make paper models, and make sure your method works for all possible triangles with any side taken as the base. In particular, make sure that your proof works for triangles whose heights are much longer than their bases. Also, you need to show that the result-ing figure actually is a parallelogram.

> **b.** *Show that, if you assume* **AP**, *then on a plane every paral-lelogram is equivalent by dissection to a rectangle with the same base and height. Show equivalence by subtraction without assuming* **AP**.

A partial proof of this was given in the introduction at the beginning of this chapter. But for this problem, your proof must also work for tall, skinny parallelograms, as shown in Figure 12.5, for which the given construction does not work. You may say that you can simply change the orientation of the parallelogram and use a long side as the base; but, as for part **a**, we want a proof that will work no matter which side you choose as the base. Be sure to note where you use **AP**. Again, do not try anything too complicated, and you only have to work on the plane.

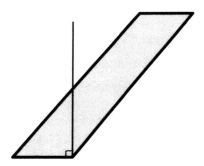

Figure 12.5   Tall, skinny parallelogram

DISSECTION THEORY ON SPHERES AND HYPERBOLIC PLANES

The above statements take on a different flavor when working on spheres and hyperbolic planes because we cannot construct parallelo-grams and rectangles, as such, on these spaces. We can define two types of polygons on spheres and hyperbolic spaces and then restate the above two problems for these spaces. The two types of polygons are the Khayyam quadrilateral and the Khayyam parallelogram. These defini-tions were first put forth by the Persian geometer-poet Omar Khayyam

(1048–1131) in the 11th century AD [**AT**: Khayyam 1958]. Through a bit of Western chauvinism, geometry books generally refer to these quadrilaterals as Saccheri quadrilaterals after the Italian priest and professor Gerolamo Saccheri (1667–1733), who translated into Latin and extended the works of Khayyam and others.

A ***Khayyam quadrilateral*** (***KQ***) is a quadrilateral such that $AB \cong CD$ and $\angle BAD \cong \angle ADC \cong \pi/2$. A ***Khayyam parallelogram*** (***KP***) is a quadrilateral such that $AB \cong CD$ and AB is a parallel transport of DC along AD. In both cases, BC is called the ***base*** and the angles at its ends are called the ***base angles***. See Figure 12.6.

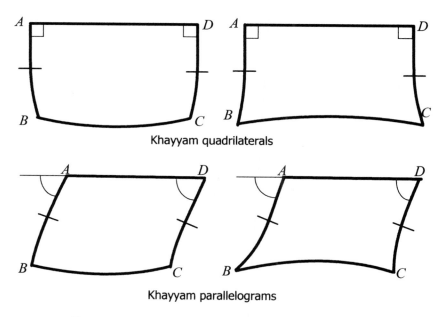

Khayyam quadrilaterals

Khayyam parallelograms

Figure 12.6   Khayyam quadrilaterals and parallelograms

PROBLEM 12.2   KHAYYAM QUADRILATERALS

a. *Prove that the base angles of a KQ are congruent.*

b. *Prove that the perpendicular bisector of the top of a KQ is also the perpendicular bisector of the base.*

c. *Show that the base angles are greater than a right angle on a sphere and less than a right angle on a hyperbolic plane.*

d. *A KQ on the plane is a rectangle and a KP on the plane is a parallelogram.*

To begin this problem, note that the definitions of KP and KQ make sense on the plane as well as on spheres and hyperbolic planes. The pictures in Figure 12.6 are deliberately drawn with a curved line for the base to emphasize the fact that, on spheres and hyperbolic planes, KPs and KQs do have the same properties as rectangles and parallelograms. You should think of these quadrilaterals and parallelograms in terms of parallel transport instead of parallel lines. Everything you have learned about parallel transport and triangles on spheres and hyperbolic planes can be helpful for this problem. Symmetry can also be useful.

Now we are prepared to modify Problem **12.1** so that it will apply to spheres and hyperbolic planes.

PROBLEM 12.3 DISSECT SPHERICAL AND HYPERBOLIC TRIANGLES AND KHAYYAM PARALLELOGRAMS

a. *Show that every hyperbolic triangle, and every small spherical triangle, is equivalent by dissection to a Khayyam parallelogram with the same base as the triangle.*

Try your proof from Problem **12.1** as a first stab at this problem. You only need to look at a sphere. You should also look at the different proofs given for Problem **12.1**. The only difference between the plane and spheres and hyperbolic planes as far as this problem is concerned is that you must be more careful on spheres and hyperbolic planes because there are no parallel lines; there is only parallel transport. Some of the proofs for Problem **12.1** work well on a sphere or on a hyperbolic plane, and others do not. Remember that the base of a KP is the side opposite the given congruent angles.

b. *Prove that every Khayyam parallelogram is equivalent by dissection (if you assume **AP**), or equivalent by subtraction* *(without assuming* **AP**)*, to a Khayyam quadrilateral with the same base.*

As with part **a,** start with your planar proof and work from there. As before, your method must work for tall, skinny KPs. Once you have come up with a construction, you must then show that the pieces actually fit together as you say they do and prove that the angles at the top are right angles. See Figure 12.7.

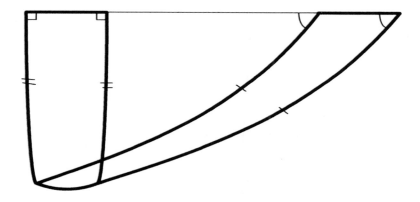

Figure 12.7   Dissecting KP into KQ

*PROBLEM **12.4** SPHERICAL POLYGONS DISSECT TO LUNES

In the next chapter you will show (under the assumption of **AP**) that every polygon on the plane is equivalent by dissection to a square, and then we will use this and the Pythagorean Theorem to show that any two polygons with the same area are equivalent by dissection. This does not apply to spheres and hyperbolic planes because there are no squares on these surfaces. However, we have already shown in Problems **7.1** and **7.4** that two polygons (or triangles) on the same sphere have the same area if they have the same holonomy. Thus, every polygon on a sphere must have the same area as some lune with the same holonomy. Now we can show that not only do they have the same area but they are also equivalent by dissection.

Assuming **AP,** *show that every simple* (sides intersect only at the vertices) *small polygon on a sphere is equivalent by dissection to a lune with the same holonomy.* That is, the angle of the lune is equal to

(½)(2π – sum of the exterior angles of the polygon).

Consequently, two simple small polygons with the same area on the same sphere are equivalent by dissection.

OUTLINE OF PROOF
The proof of this result can be completed by proving the following steps (or lemmas). (This proof was first suggested to David by his daughter, Becky, now Rebecca Wynne.)

1. *Every simple small polygon can be dissected into a finite number of small triangles, such that the holonomy of the polygon is the sum of the holonomies of the triangles.*

See Problem **7.5,** but what is needed here is easier than **7.5.**

2. *Each small triangle is equivalent by dissection to a KQ with the same base and same holonomy.*

Check your solutions for Problems **12.2** and **12.3.**

3. *Two KQs with the same base and the same holonomy (or base angles) are congruent.*

Match up the bases and see what you get.

4. *If two* Δ*s have the same base and the same holonomy, then they are equivalent by dissection.*

Put together the previous steps.

5. *Any* Δ *is equivalent by dissection to a lune with* $\mathcal{H}(\Delta) = \mathcal{H}$(lune) *= (twice the angle of the lune).*

Hint: A lune can also be considered as a triangle.

6. *Two simple small polygons on a sphere with the same area are equivalent by dissection to the same lune and therefore are equivalent by dissection to each other.*

What is the union of two lunes?

The first four steps above will also work (with essentially the same proofs) on a hyperbolic plane. But there is no clear replacement for the biangles (which do not exist on a hyperbolic plane). There is a proof of the following:

THEOREM 12.4. *On a hyperbolic space, two simple* (the sides intersect only at the vertices) *polygons with the same area are equivalent by dissection.*

Two published proofs in English are in [**TX**: Millman & Parker], page 267, and [**DI**: Boltyanski 1978], page 62. These proofs are similar, and both use the first four steps above and use the completeness of the real numbers (in the form of a version of the Intermediate Value Theorem). You can check that Becky's proof above does not use completeness. In addition, the proof of the same result on the plane (see the discussion between Problems **13.2** and **13.3**) also does not need the use of completeness axiom.

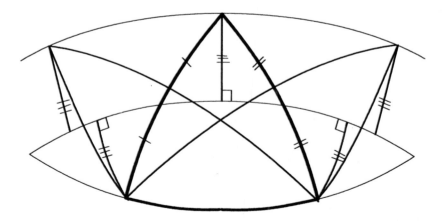

Figure 12.8   Triangles with same base and same area

In the plane all the triangles with the same base and the same height have the same area and the vertices opposite the base of these triangles form a (straight) line not intersecting the line determined by the base. On a sphere, the situation is different. Your proof above should show that midpoints of the (non-base) edges lie on a great circle and the vertices opposite the base must lie on a curve equidistant from this great circle. See Figure 12.8.

Chapter 13

SQUARE ROOTS, PYTHAGORAS, AND SIMILAR TRIANGLES

The diagonal of an oblong produces by itself both the areas
which the two sides of the oblong produce separately.
— Baudhayana, *Sulbasutram*, Sutra 48 [**AT**: Baudhayana]

This chapter is devoted to a subject that is not only of great theoretical
interest but also of recreational interest, for dissection theory plays an
important role in a rigorous development of area and volume and also
furnishes a seemingly endless variety of attractive and challenging
puzzles.

In the last chapter, we showed that two polygons on a sphere with
the same area are equivalent by dissection because both are equivalent to
the same biangle. In this chapter, we will prove an analogous result on
the plane by showing that every planar polygon is equivalent by dissection to a square. This result is the basis for many popular dissection
puzzles. Before reading this chapter you should go through the Introduction to Dissection Theory and Problem **12.1** at the start of Chapter 12.

In the process of exploring this dissection theory, we will follow a
path through a corner of the forest of mathematics — a path that has
delighted and surprised the authors many times. We will bring with us
the question, What are square roots? Along the way we will confront
relationships between geometry and algebra of real numbers, in addition
to similar triangles, the Pythagorean Theorem (the quote above is a statement of this theorem written before Pythagoras), and possibly the oldest
written proof in geometry (at least 2600 years old). This path will lead to

177

the solutions of quadratic and cubic equations in Chapter 19. We will let David's personal experience start us on this path.

SQUARE ROOTS

When David was in eighth grade, he asked his teacher, "What is a square root?" He knew that the square root of N was a number whose square was equal to N, but where would one find it? (Hidden in that question is "How does one know it always exists?") He knew what the square roots of 4 and 9 were — no problem there. He even knew that the square root of 2 was the length of the diagonal of a unit square, but what of the square root of 2.5 or of π?

At first, the teacher showed David a square root table (a table of numerical square roots), but he soon discovered that if one took the number listed in the table as the square root of 2 and squared it, you got 1.999396, not 2. (Modern-day pocket calculators give rise to the same problem but hide it reasonably effectively by calculating more digits than they display and then rounding off.) So David persisted in asking his question, What is the square root? Then the teacher answered by giving him THE ANSWER — the Square Root Algorithm. Do you re-member the Square Root Algorithm — that procedure, similar to long division, by which it is possible to calculate the square root? Or perhaps more recently you were taught the "divide and average" method, a more efficient method that was known to Archytas of Tarentum (428–350 B.C., Greek) but today is often called *Newton's Method* after Isaac Newton (English, 1643–1727).

If A_1 is an approximation of the square root of N, then the average of A_1 and N/A_1 is an even better approximation, which we could call A_2. And then the next approximation A_3 is the average of A_2 and N/A_2. In equation form this becomes

$$A_{n+1} = (1/2)(A_n+(N/A_n)).$$

For example, if $A_1 = 1.5$ is an approximation of the square root of 2, then

$$A_2 = 1.417\cdots, \quad A_3 = 1.414216\cdots$$

and so forth are better and better approximations.

But wait! Most of the time these algorithms do not calculate the square root — they only calculate approximations to the square root. The algorithms have an advantage over the tables because we could, at least in theory, calculate approximations as close as we wished. However, they are still only approximations and David's question still remained — What is the square root these algorithms approximate?

David's eighth-grade teacher then gave up, but later in college David found out that some modern mathematicians answer his question in the following way: "We make an assumption (the Completeness Axiom) that implies that the sequence of approximations from the Square Root Algorithm must converge to some real number." And, when he continued to ask his question, he found that in modern mathematics the square root is a certain equivalence class of Cauchy (Augustin-Louis Cauchy, 1789–1857, French) sequences of rational numbers, or a certain Dedekind (Julius Dedekind, 1831–1916, German) cut. Finally, David let go of the question and forgot it in the turmoil of graduate school, writing his thesis, and beginning his mathematical career.

Later, David started teaching the geometry course that is the basis for this book. One of the problems in the course was the following problem. [For this chapter you will need, at times, to assume the Archimedean Postulate, AP — when you do, note it.]

PROBLEM 13.1   A RECTANGLE DISSECTS INTO A SQUARE

Show that, on the plane, every rectangle is equivalent by dissection to a square.

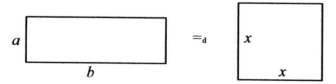

Figure 13.1   Dissecting a rectangle into a square

SUGGESTIONS

Problems 13.1 and 13.2 can be done in any order. So if you get stuck on one problem, you can still go on to the other. In Problems 13.1 and 13.2, it is especially important to make accurate models and constructions —

rough drawings will not show the necessary length and angle relation-ships.

Problem **13.1** is one of the oldest problems in geometry, so you may have guessed (correctly!) that it is one of the more complex ones. This problem is interesting for more than just historical reasons. You are asked to prove that you can cut up any rectangle (into a finite number of pieces) and rearrange the pieces to form a square, as in Figure 13.1.

Because you are neither adding anything to the rectangle nor re-moving anything, the area must remain the same, so $ab = x^2$, or $x = \sqrt{ab}$. What you are really finding is a geometric interpretation of a square root. This was done already by Pythagoreans (6th century B.C.), who called x the geometric mean of a and b.

Let us look at a proof that is similar to the proofs in many standard geometry textbooks:

Let $s = \sqrt{ab}$ be the side of the square equivalent by dissection to the rectangle with sides a and b. Place the square, *AEFH*, on the rectangle, *ABCD*, as shown in Figure 13.2. Draw *ED* to intersect *BC* in *R* and *HF* in *K*. Let *BC* intersect *HF* in *G*. [In the case that *ABCD* is so long and skinny that *K* ends up between *G* and *F*, we can, by cutting *ABCD* in half and stack-ing the halves, reduce the proof to the case above. Note that **AP** is implicitly being used here.] From the similar trian-gles $\triangle$ *KDH* and $\triangle$ *EDA* we have *HK/AE* = *HD/AD*, or

$$HK = (AE)(HD)/AD = s(a - s)/a = s - s^2/a = s - b.$$

Therefore, we have $\triangle$ *EFK* $\cong$ $\triangle$ *RCD*, $\triangle$ *EBR* $\cong$ $\triangle$ *KHD*.

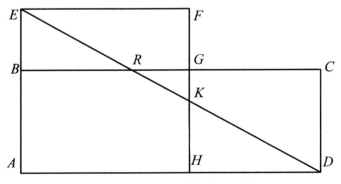

Figure 13.2   Textbook proof

David was satisfied with this proof until, in the second year of teaching the geometry course, he started sensing student uneasiness with it. As he listened to their comments, he noticed questions being asked by the students: "What is $\sqrt{ab}$?" "How do you find it?" Those used to be his questions!

The students and David also noticed that the facts used about similar triangles in the proof above are usually proved using the theory of areas of triangles. Thus, the proof could not be used as part of a concrete theory of areas of polygons, which was our purpose in studying dissection theory in the first place. Notice that the above proof also assumes the existence of the square, which in analysis is based on what is called the Completeness Axiom. The conclusion here seems to be that it would be desirable instead to construct the square root x. That started David on an exploration that continued on and off for many years.

Now let us solve a few problems in dissection theory.

Here are three methods for constructing x. For all three constructions we will use a rectangle like the one shown in Figure 13.1, with the longer side b as the base and the shorter side a as the height.

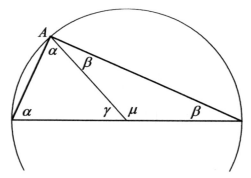

Figure 13.3   The angle at A is a right angle

For both the first and second constructions, you need to know that a triangle with all vertices on the circumference of a circle and one side a diameter of the circle is a right triangle. See Figure 13.3. That the angle not on the diameter is a right angle follows from a more general result, which we will prove in Problem **14.1a**, but the special case we need is easy to see now. In particular, draw the line segment from the vertex to the center of the circle and note that two isosceles triangles are formed. By ITT, Problem **6.2**, there are congruent angles as marked; and, because the sum of the angles in a planar triangle is 180° (Problems **7.3b** or **10.2**) we can compute that

$$(2\alpha + \gamma) + (2\beta + \mu) = 360°$$

and, because $\gamma + \mu = 180°$, it follows immediately that $\alpha + \beta = 90°$. See Problem 15.1 for a more general result.

For the first construction, Figure 13.4, take the rectangle and lay a out to the left of b. Use this base line as the diameter of a circle. The length x that you are looking for is the perpendicular line from the left side of the rectangle to where it intersects the circle.

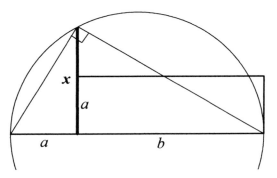

Figure 13.4   First construction of x

The second construction, Figure 13.5, is similar, but this time put a on the inside of the base of the rectangle. Now the side x you are looking for is the segment from the lower left corner of the rectangle to the point of intersection on the circle with the perpendicular rising from the point a in from the corner as indicated in the figure.

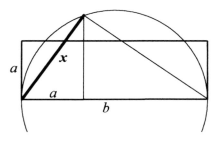

Figure 13.5   Second construction of x

The third construction, Figure 13.6, is a bit more algebraic than the others, and doesn't directly involve a circle.

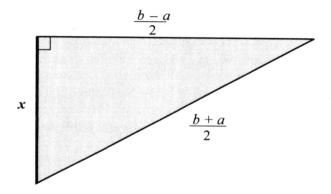

Figure 13.6   Third construction of x

This construction can be used together with the result of Problem **13.2** in order to obtain a proof by subtraction.

For all of these constructions, it is imperative that you use accurate models. Whatever method you choose, make the rectangle and the square overlap as much as possible, and see how to fit the other pieces in. Then you have to prove that all of the sides and angles line up properly. Note that it is much better to solve this problem geometrically rather than by only trying to work out the algebra — actually do the construction and proceed from there. Finally, you don't have to use one of the constructions shown here. If these don't make sense to you, then find one of your own that does.

And remember not to use results about similar triangles because we normally need results about areas in order to prove these results as we will do in Problem **13.4**.

Pause, explore, and write out your ideas before reading further.

Baudhayana's Sulbasutram

While reading an article, David ran across an item that said that the problem of changing a rectangle into a square appeared in the *Sulbasutram*. "Sulbasutram" means "rules of the cord," and the several Sanskrit texts collectively called the *Sulbasutra* were written by the Vedic Hindus starting before 600 B.C. and are thought to be compilations of oral wisdom that may go back to 2000 B.C. (See, for example, A. Seidenberg *The Ritual Origin of Geometry* [**HI**: Seidenberg].) These texts have prescriptions for building fire altars, or *Agni*. However, contained in the *Sulbasutra* are sections that constitute a geometry textbook detailing the geometry necessary for designing and constructing the altars. As far as we have been able to determine, these are the oldest geometry (or even mathematics) textbooks in existence. There are at least four versions of the *Sulbasutram* by Baudhayana, Apastamba, Katyayana, and Manava. The geometric descriptions are very similar in these four books, and we will only use Baudhayana's version here (see [**AT**: Baudhayana]).

The first chapter of Baudhayana's *Sulbasutram* contains geometric statements called "Sutra." Sutra 54 is what we asked you to prove in Problem **13.1**. It states the following:

> If you wish to turn an oblong [here "oblong" means "rectangle"] into a square, take the shorter side of the oblong for the side of square. Divide the remainder into two parts and inverting join those two parts to two sides of the square. Fill the empty place by adding a piece. It has been taught how to deduct it. [See Figure 13.7.]

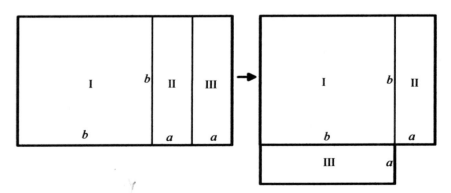

Figure 13.7   Sutra 54

So our rectangle has been changed into a figure with an "empty place," which can be filled "by adding a piece" (a small square). The result is a large square from which a small square has to be "deducted."

Now Sutra 51 (Refer to Figure 13.8.):

> If you wish to deduct one square from another square, cut off a piece from the larger square by making a mark on the ground with the side of the smaller square which you wish to deduct; draw one of the sides across the oblong so that it touches the other side; by this line which has been cut off the small square is deducted from the large one.

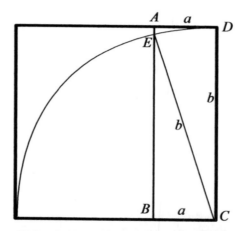

Figure 13.8   Sutra 51: Construction of side of square

We wish to deduct the small square (a^2) from the large square. Sutra 51 tells us to "scratch up" with the side of the smaller square — this produces the line AB and the oblong $ABCD$. Now, if we "draw" the side CD of the large square to produce an arc, then this arc intersects the other side at the point E. The sutra then claims that BE is side of the desired square whose area equals the area of the large square minus the area of the small square. This last assertion follows from Sutra 50, which we ask you to prove in Problem **13.2**.

See Figure 13.9 for the drawing that goes with Sutra 50:

> If you wish to combine two squares of different size into one, scratch up with the side of the smaller square a piece cut off from the larger one. The diagonal of this cutoff piece is the side of the combined squares.

Be sure you see why Sutra 50 is a statement of what we call the Pythagorean Theorem.

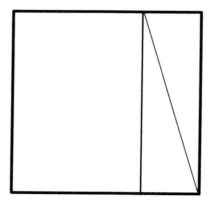

Figure 13.9   Sutra 50

A. Seidenberg, in an article entitled *The Ritual Origin of Geometry* [**HI**: Seidenberg], gives a detailed discussion of the significance of Baudhayana's *Sulbasutram*. He argues that it was written before 600 B.C. (Pythagoras lived about 500 B.C. and Euclid about 300 B.C.). He gives evidence to support his claim that it contains codification of knowledge going "far back of 1700 B.C." and that knowledge of this kind was the common source of Indian, Egyptian, Babylonian, and Greek mathematics. Together Sutras 50, 51, and 54 describe a construction of a square with the same area as a given rectangle (oblong) and a proof (based on the Pythagorean Theorem) that this construction is correct. You can find stated in many books and articles that the ancient Hindus, in general, and the *Sulbasutram*, in particular, did not have proofs or demonstrations, or they are dismissed as being "rare." However, there are several sutras in [**AT**: Baudhayana] similar to the ones discussed above. We suggest you decide for yourself to what extent they constitute proofs or demonstrations.

Baudhayana avoids the Completeness Axiom by giving an explicit construction of the side of the square. The construction can be summarized in Figure 13.10.

This is the same as Euclid's construction in Proposition II-14 (see [**AT**: Euclid, *Elements*], page 409). But Euclid's proof is much more complicated. Note that neither Baudhayana nor Euclid gives a proof of Problem **13.1** because the use of the Pythagorean Theorem obscures the dissection. However, they do give a concrete construction and a proof that the construction works.

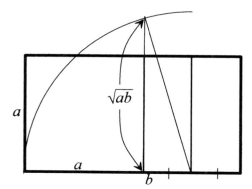

Figure 13.10   Baudhayana's construction of square root

Both Baudhayana and Euclid prove the following theorem, which uses equivalence by subtraction.

THEOREM. *For every rectangle R there are squares S_1 and S_2 such that $R + S_2$ is equivalent by dissection to $S_1 + S_2$ and thus R and S_1 have the same area.*

Notice that both Baudhayana's and Euclid's proofs of this theorem and your proof of Problem **13.1** avoid assuming that the square root exists (and, thus, avoid the Completeness Axiom). They also avoid using any facts about similar triangles. These proofs explicitly construct the square and show in an elementary way that its area is the same as the area of the rectangle. There is no need for the area or the sides of the rectangle to be expressed in numbers. Also given a real number, *b*, the square root of *b* can be constructed by using a rectangle with sides *b* and 1. In Problem **13.3** we will use these techniques to prove basic properties about similar triangles.

So, finally, we have *an* answer to our question, What is a square root? Also, see David's chapter, "Square Roots in the Sulba Sutra," in the book [**EG**: Gorini], which describes a geometric method based on Baudhayana that finds arbitrarily accurate numerical approximations to many square roots and does so in a way that is computationally more efficient than the "divide and average" method (*Archytas/Newton's Method*) described at the beginning of this chapter.

PROBLEM 13.2   EQUIVALENCE OF SQUARES

Prove the following: On the plane, the union of two squares is equivalent by dissection to another square.

SUGGESTIONS

This is closely related to the Pythagorean Theorem. There are two general ways to approach this problem: You can use Problem **13.1** or you can prove it on its own, which will result in a proof of the Pythagorean Theorem — hence, you can't use the Pythagorean Theorem to solve this problem because you will be proving it!

To see how this problem relates to the Pythagorean Theorem, think about the following statement of the Pythagorean Theorem:

The square on the hypotenuse is equal to the sum of the squares on the other two sides.

This is not just an algebraic equation — **the squares referred to are actual geometric squares**.

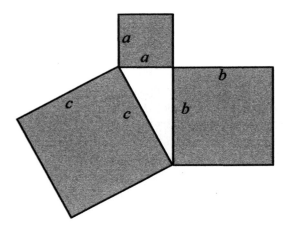

Figure 13.11   Pythagorean Theorem

As in the other dissection problems, make the three squares coincide as much as possible, and see how you might get the remaining pieces to overlap. You might start by reflecting the square with side c over the side c on the triangle. Then prove that the construction works as you say it does.

ANY POLYGON CAN BE DISSECTED INTO A SQUARE

If we put together Problems **12.1**, **13.1**, and **13.2**, we get the surprising

>   **THEOREM 13.2a** *On the plane, every simple polygon is equivalent by dissection to a square.*

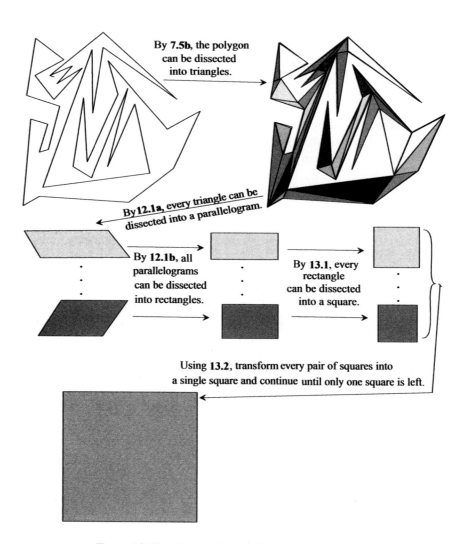

Figure 13.12   Every polygon dissects into a square

HISTORY OF DISSECTIONS

We have now seen dissection proofs in Chapter 12, Problems **13.1** and **13.2**, and we will see more in later chapters (for example, the dissection proof of the Law of Cosines in **20.2a**). In the section History of Dissections in the Theory of Area of Chapter 12 we outlined how dissection have been used to prove results about area and to develop a rigorous theory of area. In this chapter we have seen more examples of this.

Note that the result in Theorem 13.2a implies, on the plane, that

> **THEOREM 13.2b** *Any two polygons with the same area are equivalent by dissection to each other.*

This theorem is true not only on the plane but also on spheres (Problem **12.4**) and hyperbolic planes (Theorem 12.4). On the plane the theorem is often called the *Bolyai-Gerwien-Wallace Theorem* (with sometimes Wallace's name left off) because apparently it was first proved by them: Wallace in 1831, Bolyai in 1832, and Gerwien in 1833.

Note that Theorem 13.2 does not tell us anything about how many pieces would be needed in any particular dissection. In fact, the number will, in general, be very large. So this leads to the question of making dissections with the least number of pieces, or with the most symmetry, or, in some other way, with the most elegance. A large number of polygonal dissection puzzles and recreations arises from these questions. There is a surprisingly rich history associated with recreational dissections. The most recent accounts of dissection puzzles and their history is in [**DI**: Frederickson 1997] and [**DI**: Frederickson 2002]. Refer to these books for further history and about 1000 dissection pictures. There are also further references in the Dissection section of the Annotated Bibliography.

As an example, the dissection in Figure 13.13 is a famous dissection of a equilateral triangle into a square that requires only four pieces. The discovery of this dissection is usually credited to Henry Dudeney (English, 1857–1930) in 1902, but [**DI**: Frederickson 1997], p. 136, raises some doubts about this attribution.

In this dissection, the segments *AD*, *DB*, *BE*, *EC*, *FG* are all equal to half the side of the triangle; *EF* is equal to the side of the equivalent square; *DJ* and *GK* are each perpendicular to *EF*. If the four pieces are successively hinged to one another at the points, *D*, *E*, *G*, then holding piece I fixed and swinging the connected set of pieces IV-III-II clock-wise, the equilateral triangle is neatly carried into the square.

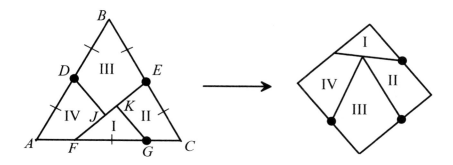

Figure 13.13 Four-piece dissection of triangle to square

We could build a set of four connected tables based on this fact; swinging the tables in one direction causes the tops to fit together into a single equilateral triangular table and swinging them in the other direction causes the tops to fit together into a single square table. This is an example of a hinged dissection. If you go back to your dissections in Chapter 12 and **13.1** and **13.2**, you will probably find that some of them can also be hinged. The [**DI**: Frederickson 2002] book is entirely about hinged dissections and what he calls "twisting dissections."

The 18th and 19th centuries produced a variety of puzzle books that contained dissection puzzles. Articles about dissections also appeared in scientific journals. In the early 19th century the Bolyai-Gerwien-Wallace Theorem was proved. Dissection puzzles had great popularity during 1880s and 1890s, and at the end of 19th century puzzle columns by Sam Loyd and Henry E. Dudeney began to appear in newspapers and magazines. The 20th century brought an increase in sophistication to dissections. Articles and problems appeared in periodicals such as *Mathematical Gazette*, *American Mathematical Monthly*, and *Scientific American*. Geometric dissection has become a frequent topic in books on mathematical recreations, but there are books that focus exclusively on dissections, [**DI**: Lindgren 1964 & 1972] and [**DI**: Boltyanski 1980].

See the references in the Dissections section of the Bibliography for more about dissections.

PROBLEM 13.3 MORE DISSECTION-RELATED PROBLEMS

Some of these dissection problems are easy and some are hard. We do not tell you which; but, in any case, what is hard for us may be easy for you.

a. *Show that the four-piece dissection of an equilateral triangle to a square depicted in Figure 13.13 actually works as claimed.*

b. *Show that any obtuse triangle* (one angle is greater than a right angle) *can be dissected into acute triangles* (all angles are less than a right angle). *What is the dissection with the smallest number of pieces you can find?*

c. *Prove that the figure formed when the midpoints of the sides of a quadrilateral are joined in order is a parallelogram, and its area is half that of the quadrangle.* Remember that quadrilaterals need not be convex; see Figure 13.14.

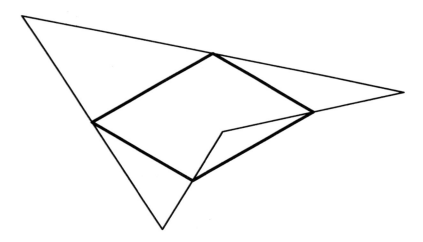

Figure 13.14   The Varignon parallelogram of a quadrilateral

This parallelogram is often referred to as the *Varignon parallelogram* of the quadrilateral, because the result first appeared in a 1731 posthumous publication by Pierre Varignon (1654–1722).

d. *Show that if one diagonal divides a quadrangle into two triangles of equal area, it bisects the other diagonal. Conversely, if one diagonal bisects the other, it bisects the area of the quadrangle.*

e. *Divide a triangle by two straight lines into three parts, which when properly arranged shall form a parallelogram whose angles are of given magnitudes.*

f. *From the arbitrary point within a triangle, draw three lines to the sides of the triangle that shall trisect its area.*

g. *Dissect a regular hexagon by straight cuts into six pieces that can be reassembled to form an equilateral triangle. Does there exist a five-piece solution?*

h. *Given $6 \geq n$, show that it is possible to fit together n isosceles right triangles, all of different sizes, so as to make a single isosceles right triangle. Can the problem be solved for $n = 5$? Can a right isosceles triangle be dissected into a finite number of right isosceles triangles, no two having a common side?*

i. *Show that any given triangle can be dissected by straight cuts into four pieces that can be arranged to form two triangles similar to the given triangle.*

j. *Dissect a regular pentagon by straight cuts into six pieces that can be put together to form an equilateral triangle.*

k. *Draw a straight line that will bisect both the area and the perimeter of a given quadrilateral.*

*THREE-DIMENSIONAL DISSECTIONS AND HILBERT'S THIRD PROBLEM

In 1900, David Hilbert delivered a lecture before the International Congress of Mathematicians in which he listed ten problems "from the discussion of which an advancement of science may be expected." In a later paper he expanded these to 23 problems, which are now called *Hilbert's Problems*.

> **Hilbert's Third Problem.** *Is it possible to specify two tetrahedra of equal bases and equal altitudes that can in no way be split up into congruent tetrahedra and that cannot be combined with congruent tetrahedra to form two polyhedra that themselves could not be split up into congruent tetrahedra?* See [**DI**: Sah], [**DI**: Boltyanski 1978], and [**TX**: Hartshorne] for a discussion of this problem and its history.

Shortly after Hilbert's lecture, Max Dehn (1878–1952, Germany, United States) found such tetrahedra and also proved that a regular tetrahedron is not equivalent by dissection to a cube. Thus there is no possibility of dissecting polyhedra into cubes. To show these results, Dehn proved the following:

> **THEOREM 13.4** *If P and Q are two polyhedra in 3-space that are equivalent by dissection (or by subtraction), then the dihedral angles (see Chapter 21) of P are, mod π, a rational linear combination of the dihedral angles of Q. That is, if α_i are the dihedral angles of P and β_j are the dihedral angles of Q, then there are integers n_i, m_j, a, b such that*
>
> $$\textstyle\sum_i n_i\alpha_i + a\pi = \sum_j m_j\beta_j + b\pi.$$

This interesting fact is of far-reaching importance in the theory of volumes. It implies that although a theory of areas of polygons can be constructed without continuity considerations, such is not possible for a complete theory of volumes of polyhedra. Thus the ordinary treatment of volumes, such as we find in Euclid's Elements, involves infinitesimals, which do not occur in the theory of polygonal areas. In addition, we showed in Problem **7.5b** that though it is always possible to dissect a polygon into a finite number of triangles having their vertices only at the vertices of the polygon, there exist polyhedra, called Lennes polyhedra, which cannot be dissected into finite numbers of tetrahedra having their vertices only at the vertices of the polyhedra.

PROBLEM **13.4**   **SIMILAR TRIANGLES**

Near the beginning of this chapter we gave a textbook proof of Problem **13.1**, which used properties of similar triangles. Later you found a proof that did not need to use similar triangles. Remember that in our discussion of AAA we said that two triangles were *similar* if their corresponding angles are congruent. Now you are ready to give a dissection proof of the following:

a. **AAA similarity criterion:** *If two triangles are similar, then the corresponding sides of the triangles are in the same proportion to one another.*

SUGGESTIONS

Look at your proof of Problem **13.1**. It probably shows implicitly that Problem **13.4** holds for a pair of similar triangles in your construction. For more generality, let θ be one of the angles of the triangles and place the two θ's in VAT position in such a way that they form two parallelograms as in Figure 13.15.

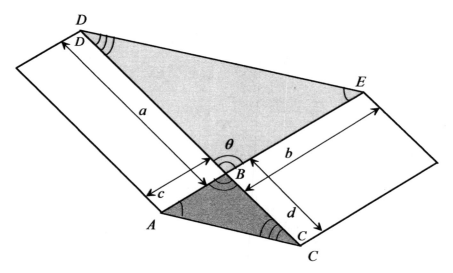

Figure 13.15   Similar triangles

Show that the two parallelograms are equivalent by dissection and use that result to show that $ac = bd$ or, in equivalent form, $a/d = c/b$. You may find it clearer if you start by looking at the special case of $\theta = \pi/2$.

b. **SAS similarity criterion:** *If two triangles on a plane have an angle in common and if the corresponding sides of the angle are in the same proportion to each other, then the triangles are similar.*

Draw the two triangles with their common angle coinciding. Then show that the opposite sides are parallel. You will probably have to use part **a**.

These results about similar triangles will be used in most of the remaining chapters of the book.

Chapter 14

PROJECTIONS OF A SPHERE ONTO A PLANE

> Geography is a representation in picture of the whole known world together with the phenomena which are contained therein.
>
> ... The task of Geography is to survey the whole in its proportions, as one would the entire head. For as in an entire painting we must first put in the larger features, and afterward those detailed features which portraits and pictures may require, giving them proportion in relation to one another so that their correct measure apart can be seen by examining them, to note whether they form the whole or a part of the picture. ... Geography looks at the position rather than the quality, noting the relation of distances everywhere, ...
>
> It is the great and the exquisite accomplishment of mathematics to show all these things to the human intelligence ...
>
> — Claudius Ptolemy, *Geographia*, Book One, Chapter I

A major problem for map makers (cartographers) since Ptolemy (approx. 85–165 A.D., Alexandria, Egypt) and before is how to represent accurately a portion of the surface of a sphere on the plane. It is the same problem we have encountered when making drawings to accompany our discussions of the geometry of the sphere. We shall use the terminology used by cartographers and differential geometers to call any one-to-one function from a portion of a sphere onto a portion of a plane a *chart*. As Ptolemy states in the quote above, we would like to represent the sphere on the plane so that proportions (and thus angles) are preserved and the relative distances are accurate. For a history and mathematical descriptions of charts of the sphere, see [**CE**: Snyder]. For a history and

197

discussion of the political, social, and ethnic controversies that have been and continue to be connected with maps and map making, see [**CE: Monmonier**]. For a discussion of how maps and the stars were used to determine the position of the earth, see the delightful book [**CE: Sobel**]. History of specific map projections is given in the last section of this chapter.

In this chapter we will study various charts for spheres. We will need properties of similar triangles that are investigated in Problem **13.3**.

PROBLEM 14.1   CHARTS MUST DISTORT

It is impossible to make a chart without some distortions.

> *Which results that you have studied so far show that there must be distortions when attempting to represent a portion of a sphere on the plane?*

Nevertheless, there are projections (charts) from a part of a sphere to the plane that do take geodesic segments to straight lines, that is, that preserve the shape of straight lines. There are other projections that preserve all areas. There are still other projections that preserve the measure of all angles. In this chapter, we will study these three types of projections on a sphere, and in Chapter 17 we will look at projections of hyperbolic planes.

PROBLEM 14.2   GNOMIC PROJECTION

Imagine a sphere resting on a horizontal plane. See Figure 14.1. A *gnomic projection* is obtained by projecting from the center of a sphere onto the plane. Note that only the lower open hemisphere is projected onto the plane; that is, if X is a point in the lower open hemisphere, then its gnomic projection is the point, $g(X)$, where the ray from the center through X intersects the plane.

> **a.** *Show that a gnomic projection takes the portions of great circles in the lower hemisphere onto straight lines in the plane. (A mapping that takes geodesic segments to geodesic segments is called a **geodesic mapping**.)*

b. *Gnomic projection is often used to make navigational charts for airplanes and ships. Why would this be appropriate?*

Hint: Start with our extrinsic definition of "great circle."

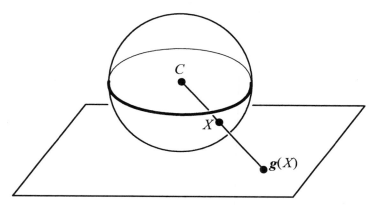

Figure 14.1   Gnomic projection

PROBLEM **14.3**   CYLINDRICAL PROJECTION

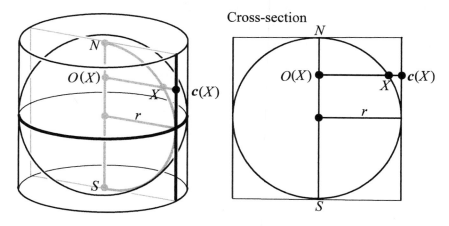

Figure 14.2   Cylindrical projection

Imagine a sphere of radius r, but this time center it in a vertical cylinder of radius r and height $2r$. The *cylindrical projection* is obtained by projecting from the axis of the cylinder, which is also a diameter of the sphere; that is, if X is a point (not the north or south poles) on the sphere and $O(X)$ is the point on the axis at the same height as X, then X is projected onto the intersection of the cylinder with the ray from $O(X)$ to X. See Figure 14.2.

a. *Show that cylindrical projection preserves areas.* (Mappings that preserve area are variously called **area-preserving** or **equiareal**.)

Geometric Approach: Look at an infinitesimal piece of area on the sphere bounded by longitudes and latitudes. Check that when it is projected onto the cylinder the horizontal dimension becomes longer but the vertical dimension becomes shorter. Do these compensate for each other?

*Analytic Approach: Find a function f from a rectangle in the (z, θ)-plane onto the sphere and a function h from the same rectangle onto the cylinder such that $c(f(z, \theta)) = h(z, \theta)$. Then use the techniques of finding surface area from vector analysis. (For two vectors A, B, the magnitude of the cross product $|A \times B|$ is the area of the parallelogram spanned by A and B. An element of surface area on the sphere can be represented by $|f_z \times f_\theta| \, dz \, d\theta$, the cross product of the partial derivatives.)

We can easily flatten the cylinder onto a plane and find its area to be $4\pi r^2$. We thus conclude the following:

b. *The (surface) area of a sphere of radius r is $4\pi r^2$.*

PROBLEM 14.4 STEREOGRAPHIC PROJECTION

Imagine the same sphere and plane, only this time project from the uppermost point (north pole) of the sphere onto the plane. This is called *stereographic projection*.

a. *Show that stereographic projection preserves the sizes of angles.* (Mappings that preserve angles are variously called **angle-preserving**, **isogonal**, or **conformal**.)

SUGGESTIONS

There are several approaches for exploring this problem. Using a purely geometric approach requires visualization but only very basic geometry.

An analytic approach requires knowledge of the differential of a function from $\mathbf{R}^2$ into $\mathbf{R}^3$. See Figure 14.3.

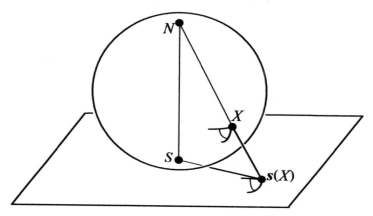

Figure 14.3   Stereographic projection is angle-preserving

Geometric Approach: An angle at a point X on the sphere is determined by two great circles intersecting at X. Look at the two planes that are determined by the north pole N and vectors tangent to the great circles at X. Notice that the intersection of these two planes with the horizontal image plane determines the image of the angle. Because the 3-dimensional figure is difficult for many of us to imagine in full detail, you may find it helpful to consider what is contained in various 2-dimensional planes. In particular, consider the plane determined by X and the north and south poles, the plane tangent to the sphere at X, and the planes tangent to the sphere at the north and south poles. Determine the relationships among these planes.

*Analytic Approach: Introduce a coordinate system and find a formula for the function s^{-1} from the plane to the sphere, which is the inverse of the stereographic projection s. Use the differential of s^{-1} to examine the effect of s^{-1} on angles. You will need to use the dot (inner) product and the fact that the differential of s^{-1} is a linear transformation from the (tangent) vectors at $s(X)$ to the tangent vectors at X.

   *b. *Show that stereographic projection takes circles through N to straight lines and circles not through N to circles.*
   (Such mappings are called ***circle-preserving***.)

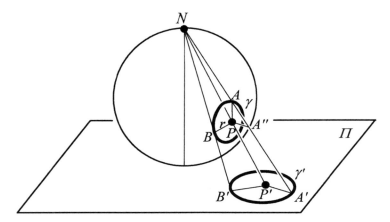

Figure 14.4   Stereographic projection is circle-preserving

SUGGESTIONS

Let γ be a circle on the sphere with points A and B and let γ', A', B' be their images under stereographic projection. Form the cone that is tangent to the sphere along the circle γ and let P be its cone point (note that P is not on the sphere). See Figure 14.4. Thus the segments BP and AP are tangent to the sphere and have the same length r. Look in the plane determined by N, A, and P and show that $\angle PAA'$ is congruent to $\angle AA'P'$. You probably have already proved this in part **a**; if not, look at the plane determined by N, A, and P and its intersections with the plane tangent to the north pole N and the image plane Π. In this plane draw line PA'' parallel to $P'A'$. Then use similar triangles and **6.2c** to show

$$|A'P'| = r\left(\frac{|NP'|}{|NP|}\right),$$

and thus γ' is a circle with center at P'.

HISTORY OF STEREOGRAPHIC PROJECTION AND ASTROLABE

The earliest known uses of projections of a sphere onto the plane were in Greece and were for purposes of map making (as described in the quote at the beginning of this chapter from Ptolemy's *Geographia*). In addition, as we discussed in Chapter 2, the Greeks (following the Babylonians) in 4th century B.C. considered the visible cosmos to be a sphere with three different coordinate systems: celestial, ecliptic, and horizon. At least as early as 2nd century B.C., the Greek and later the Arab mathematicians used the sphere projections that we have studied to represent these three spheres on the a plane.

The earliest references to stereographic projection in literature are given by Vitruvius (Roman, ~100 B.C.), *Ten Books on Architecture,* and in Ptolemy's *Representation of the Sphere in the Plane.* The ancients knew how to prove (using propositions from Apollonius' *Conics* [**AT:** Appollonius]) some properties of stereographic projection, such as the following: Circles through the pole are mapped onto straight lines and all other circles are mapped onto circles. (See Problem **14.4b**.)

In Ptolemy's work the description of stereographic projection was used for a horoscopic instrument for determining time. Later the word *horoscope* denoted the point of intersection of the ecliptic and the eastern part of the horizon determined by means of this instrument. Probably Theon of Alexandria (Greek, ?335–?405) was the first to combine stereographic projections of the three-sphere model onto a single compact planar instrument called an ***astrolabe***, as pictured in Figure 13.15. Theon's work has not survived, but we know of it because of two surviving works on the astrolabe: *On the Construction and Use of the Astrolabe* by Philoponus (6th century Alexandria) and the *Treatise on the Astrolabe* by Severus Sebokt (7th century Syria).

The astrolabe was widely known in the medieval East and Europe from the 9th century until the 19th century and used to solve problems concerning the apparent positions of the stars, sun, moon, and planets for use in navigation, time-telling, and astrology. In the Middle Ages stereographic projection was often called "astrolabe projection." The astrolabe allowed for determining relative positions to about 1° accuracy.

The term "stereographic projection" was apparently first introduced by Francois D'Aguillon (1566–1617) in his *Six Books of Optics.* The earliest exposition of the theory of stereographic projection with proofs was the *Book on the Construction of the Astrolabe* by the 9th-century Baghdad scholar al-Farghani. Mathematicians in the medieval East also tried to use other geometric transformations for constructing astrolabes. The 10th-century scholar al-Saghani suggested a projection from an arbitrary point on the axis — if that point is the center of the sphere, then this projection is the gnomic projection of Problem **14.2**. In such a projection circles on the sphere are mapped onto conics. Al-Biruni (973–1048, now Uzbekistan) described several ways of constructing an astrolabe; including the use a cylindrical projection (Problem **14.3**).

Figure 14.5 Astrolabe

In the Figure 14.5 the concentric circles on the background plate about the center represent stereographic projection of the celestial sphere from its south pole — the center of the astrolabe represents the north star, the outer rim represents the southern tropic (the furthest the sun goes south, about –24°), and the concentric circle about two-thirds of the way out is the celestial equator. The metal annulus that is off-center is the stereographic projection of the ecliptic (the path of the sun and planets) — this projection is a circle but the equal divisions of the ecliptic (the 360 degrees of the ecliptic) are not projected to equal arcs. The tight system of circles on the background plate of the astrolabe (and mostly above the center) is the projection of the coordinates of the visible hemisphere with the horizon (not complete) on the outside and the image of the zenith (the point in the sky directly over head) where the coordinate circles converge. Since the relationship between the North Star and the zenith changes with latitude, a given astrolabe is only accurate at one latitude (about 46° north for the one in the photo).

For more about astrolabes and map making, see [**HI**: Rosenfeld], pp. 121–130, [**HI**: Evans], pp. 141–162, and [**HI**: Berggren], pp. 165–186.

Chapter 15

CIRCLES

... the Power of the World always works in circles, and everything tries to be round.
— Black Elk in *Black Elk Speaks* [**GC**: Neihardt]

We now study some important properties (Problem **15.1**) of circles in the
plane that are stated and proved in Euclid's *Elements*. These planar
results will be used in later chapters, and they are also often studied for
their own interest. In Problem **15.2**, we will explore a recent (June 2003)
extension of these results to circles on spheres (and later to hyperbolic
planes). We will end the chapter with applications of these properties of
circles (Problem **15.3**) to the ancient problem of trisecting angles
(Problem **15.4**).

For Chapter 15, the only results needed from Chapters 9–14 are

PROBLEM 13.4a: The **AAA similarity criterion** for triangles
on the (Euclidean) plane: *If two triangles are similar* (have
congruent angles), *then the corresponding sides of the triangles are in the same proportion to one another.* [Needed
throughout this and later chapters.]

PROBLEM 9.1: **Side-Side-Side**: *If two triangles* (small triangles
if on a sphere) *have congruent corresponding sides, then the
triangles are congruent.* [Needed throughout this and later
chapters.]

PROBLEM 14.4: *Stereographic projection of a sphere onto a
plane preserves angles, takes circles to circles (or to straight
lines).* [Needed only for Problem **15.2**.]

If you are willing to assume these results, then you can work through Chapter 15 without Chapters 9–14.

PROBLEM 15.1   ANGLES AND POWER OF POINTS
FOR CIRCLES IN THE PLANE

These results are all stated and proved in Euclid's *Elements*. They are contained in Propositions 27, 32, and 35–37 of Euclid's Book III, which is entirely devoted to properties of planar circles.

> **a.**   *If an arc of a circle subtends an angle 2α from the center of the circle, then the same arc subtends an angle α from any point on the circumference. In particular, two angles that subtend (from different points on the circumference) the same arc are congruent.*

Use Figures 15.1 and 15.2. Draw a segment from the center of the circle to the point *A* and use ITT. Note the four different locations for *A*.

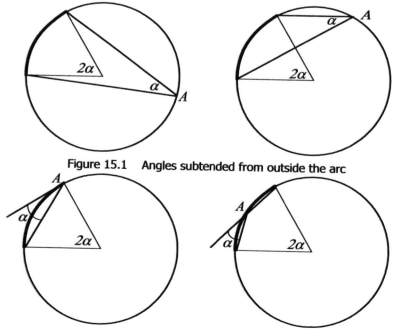

Figure 15.1   Angles subtended from outside the arc

Figure 15.2   Angles subtended from on the arc

b. *On a plane, if two lines through a point P intersect a circle at points A, A' (possibly coincident) and B, B' (possibly coincident), then*

$$|PA| \times |PA'| = |PB| \times |PB'|.$$

*This product is called the **power of the point P with respect to the circle**.*

Use Figures 15.3 and 15.4 and draw the segment joining A to B' and the segment joining A' to B. Then apply part **a** and look for similar triangles.

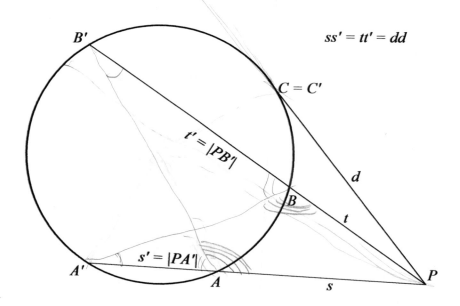

Figure 15.3   Power of a point outside with respect to a circle

For more discussion on power of a point on the plane, see the delightful little book [**EG**: Coxeter & Greitzer].

In Problem **15.2** we will state and prove results about the power of a point on spheres and hyperbolic planes that were discovered by Robin Hartshorne and published in 2003.

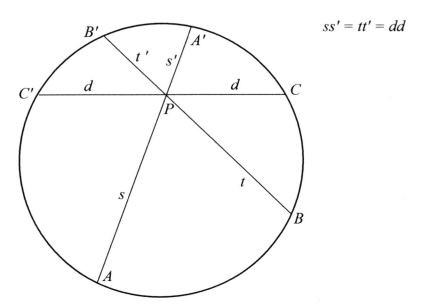

$$ss' = tt' = dd$$

Figure 15.4   Power of a point inside with respect to a circle

*PROBLEM 15.2 POWER OF POINTS
FOR CIRCLES ON SPHERES

There are no similar triangles on spheres and hyperbolic planes, so it seems surprising that there could be a notion of power of a point for circles on spheres or hyperbolic planes. However, Robin Hartshorne, [SP: Hartshorne], recently found a way to provide a unified definition of power of a point for circles in non-Euclidean geometries — this problem is based on his paper. Hartshorne starts by pointing out that the equality of the products $|PA| \times |PA'| = |PB| \times |PB'|$ in **15.1b** that we used to define the power of a point for plane circles can be written as an equality of areas of the rectangles with sides, *PA*, *PA'*, and *PB*, *PB'*, respectively — this is the way Euclid considered the power of a point. Of course, there are no rectangles on spheres and hyperbolic planes. What can we use in their place? In Chapter 12, for our work in dissection theory, we introduced Khayyam quadrilaterals as an appropriate analogue of rectangles on spheres and hyperbolic planes. Hartshorne instead introduces another analogue of a rectangle: A *semi-rectangle* is a quadrilateral with opposite sides equal and at least one right angle.

a. *On the plane, semi-rectangles are rectangles. On spheres and hyperbolic planes, semi-rectangles have the angle opposite the right angle also right and the two remaining angles are congruent.* See Figure 15.5.

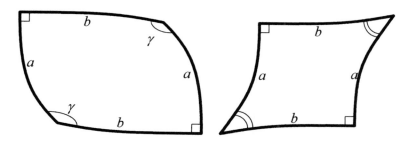

Figure 15.5 Semi-rectangles

There is a close relationship between semi-rectangles and Khayyam quadrilaterals. In fact, the interested reader can show that a semi-rectangle can be dissected, with only one straight cut, into a Khayyam quadrilateral whose base angles are the non-right angles in the semi-rectangle.

Using semi-rectangles, we can now restate **15.1b** as follows:

THEOREM 15.2 *On a plane, sphere, or hyperbolic plane, if two lines through a point P intersect a circle at points A, A' (possibly coincident) and B, B' (possibly coincident), then the area of the semi-rectangle with side lengths PA and PA' is equal to the area of the semi-rectangle with side lengths PB and PB'.* This area is called the ***power of the point P with respect to the circle***.

On the plane this theorem reduces directly to **15.1b**. Do you see why? We will now prove Theorem 15.2 on spheres (part **c**), but first we need to prove the following result, which is interesting in its own right:

b. *On a sphere with radius ρ, a spherical right triangle with excess δ and legs* (sides adjacent to the right angle) *a and b satisfies the following*:

$$\tan(\delta/2) = \tan(a/2\rho)\tan(b/2\rho).$$

Thus, if R is a semi-rectangle with sides a and b and with non-right angles equal to γ, as in Figure 15.5, and if we set let δ = γ − π/2, then

$$\text{Area}(R) = 2\delta\rho^2 \quad and \quad \tan(\delta/2) = \tan(a/2\rho)\tan(b/2\rho).$$

The proofs of parts **b** and **c** (below) rely on properties of stereographic projection which is discussed in Problem **14.4**. The properties we need here are that stereographic projection preserves all angles and takes circles on the sphere to circles (or lines) on the plane.

OUTLINE OF PROOF OF 15.2B:

1. Rotate the sphere until the right angled vertex on the triangle is at the south pole, S. Now, using stereographic projection from the north pole, project Δ onto the plane. The image of the triangle on the plane is the figure SHI depicted in Figure 15.6. The sides SH and SI are straight. (*Why?*) Let Λ be the image of the great circle that contains the hypotenuse of Δ and let C be the center of Λ. Let Γ be the image of the equator; then Γ is a circle with center S and radius 2ρ. (*Why?*) The circle Λ (that contains the arc HI) intersects the circle Γ at diametrically opposite points DE. (*Why?*)

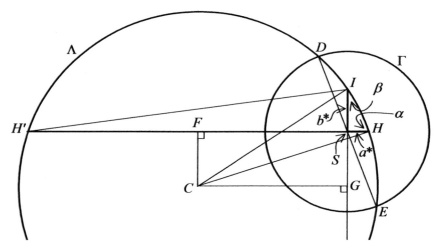

Figure 15.6 Stereographic image of a spherical right triangle

2. Referring to Figure 15.6, show that $\angle ICH = \delta$ and that $\angle IH'H = \delta/2$. (Hint: Show that $\angle FCI = \pi/2 - \beta$.)

3. Show that $|SH| \times |SH'| = |SD| \times |SE|$ and thus $|SH'| = 4\rho^2/a^*$. Conclude that $4\rho^2 \tan(\delta/2) = a^*b^*$. Remember that a^* and b^* are the projections of the sides of our original semi-rectangle R and that $\delta\rho^2$ is the area of the triangle.

4. Let R be a semi-rectangle (as in Figure 15.5) whose non-right angles are both γ, then use a diagonal to divide R into two congruent right triangles with legs a and b and non-right angles α and β. Note that $\alpha+\beta = \gamma$. This is enough to conclude (*Why?*) the last equations in part **b**.

Now we are ready to

c. *Prove that Theorem 15.2 is true on a sphere.*

OUTLINE OF PROOF OF **15.2c**

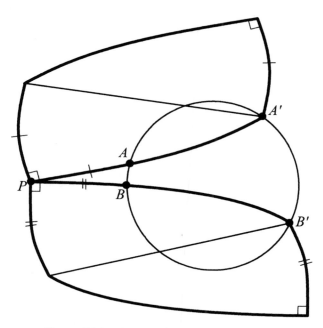

Figure 15.7   Power of a point on a sphere

1. Now let P, A, A', B, B' be as in Theorem 15.2 and Figure 15.7. By rotating the sphere, if necessary, we may consider P to be the south pole S. Let Γ denote the given circle on sphere. Project stereographically everything onto the plane and let A^*, A'^*, B^*, B'^*, and Γ^* denote the corresponding images on the plane after stereographic projection. Note that the image of $P = S$ is the same S and that A^*-S-A'^*, and B^*-S-B'^* are straight lines. (*Why?*) Using **15.1b**, conclude that

$$|SA^*| \times |SA'^*| = |SB^*| \times |SB'^*|.$$

2. Putting this together with Steps 3 and 4 of the proof of **15.2b**, we conclude (*Why?*) that

$$\tan(\delta/2) = \tan(\delta'/2),$$

where the area of the semi-rectangle on PA and PA' is $2\delta\rho^2$ and the area of the semi-rectangle on PB and PB' is $2\delta'\rho^2$. See Step 1. Thus the areas are equal. (*Why?*)

There is a proof of Theorem 15.3 on a hyperbolic plane that is very similar to the proof above on a sphere. Instead of stereographic projection, the proof on a hyperbolic plane uses the Poincaré disk model thought of as a projection (see Problem **17.5**), which also preserves angles and takes circles on the hyperbolic plane to circles (or lines) on the plane.

PROBLEM 15.3 APPLICATIONS OF POWER OF A POINT

Here are a few applications of the notion of power of a point that hold on the plane, spheres, and hyperbolic planes. We will meet other applications later, especially in Chapter 16 and 19, but these will be applied only in the case of the Euclidean plane. The applications here relate to the notion of **radical axis of two circles**, which we define to be the locus of points P such that the power of the point P is the same with respect to both circles. You may assume that Theorem 15.2 is true on hyperbolic planes.

  a. *If two circles intersect in two points, then their radical axis is the full line determined by the two points of intersection. If two circles are tangent, then the radical axis is their common tangent line.*

b. *If P is a point on the radical axis of two circles, C and D, then any circle with center P that intersects C at right angles also intersects D at right angles.*

c. *If three circles intersect each of the other two circles in two points, then the three chords so defined intersect in a common point. See Figure 15.8.*

d. *What happens if, in part **c**, some of the pairs of circles are tangent instead of intersecting in two points?*

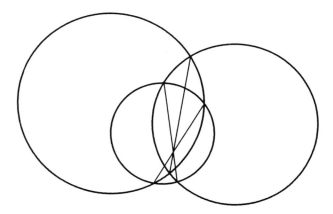

Figure 15.8 Intersecting chords

PROBLEM **15.4** TRISECTING ANGLES AND OTHER CONSTRUCTIONS

In Problem **6.3** you showed how to bisect angles and find the perpendicular bisector of a line segment by only using a compass (for drawing circles) and an unmarked straightedge (for drawing line segments joining two points but not used for measuring). We will now extend these constructions and discuss in what sense angles can and cannot be trisected.

    a. *Show, using a compass and unmarked straightedge, how to*
        i. *Construct a line from a given point perpendicular to a given line.*
        ii. *Construct the line tangent to a given circle at a given point.*
        iii. *Construct the two lines tangent to a given circle from a given point outside the circle.* [Hint: Use **15.1b.**]
        iv. *Construct a line that is a parallel transport of a given line along a given transversal.*

    b. *On the plane, show how to n-sect a given line segment using only a compass and (unmarked) straightedge. Will your construction work on spheres or hyperbolic planes?*

Hint: Look at Figure 15.9, where the given segment is *AB* and AC_1 is any segment forming an acute angle with *AB*. Duplicate AC_1 *n* times. (*How?*) Draw a line through C_1 parallel to BC_n. In the figure, *n* equals 5.

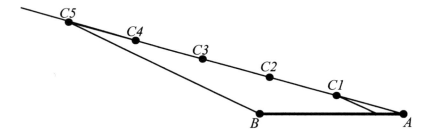

Figure 15.9 Dividing a given line segment into 5 congruent pieces

It is often stated in popular literature that it is impossible to trisect angles with straightedge and compass. See [**TX**: Martin], page 49. However, it has been known since ancient Greek times that any angle can be trisected using only a (marked) straightedge and compass. For example, Archimedes (287–212 B.C.) showed how to trisect any angle (less than 135°) using a ***marked straightedge*** (a straightedge with two points marked on it).

    c. **Archimedes Construction** (for angles less than 90°): *Referring to Figure 15.10, let ∠ABC be an angle less than 90°. Assume you have a straightedge with two marks on it that are a distance r apart. Draw the circle of radius r with center B and, keeping* *the angle the same, move A to the intersection of BA with the circle. Lay the straightedge on the figure so that it passes through A; and one mark, D, is on the circle and the other mark, E, is on the extension of BC. Show that ∠AEC is 1/3 of ∠ABC.*

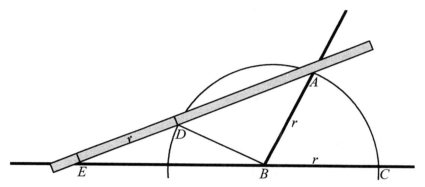

Figure 15.10 Archimedes' construction: ∠*AEC* is 1/3 of ∠*ABC*

d. *Show that the mechanism in Figure 15.11 will trisect an angle. Will it work on spheres?*

Figure 15.11 Mechanism to trisect an angle

OK, so then maybe it is impossible to trisect angles with a compass and an *unmarked* straightedge?

e.   *Construct on a transparent sheet the figure on the left in Figure 15.12 with compass and unmarked straightedge starting with any two points AB. Lay the transparent sheet on the angle α (see the figure on the right in Figure 15.12) so the vertex, V, of the angle lies on DB, one side contains C, and the other is tangent to the semicircle with center A. Prove that ∠BVC trisects angle α.*

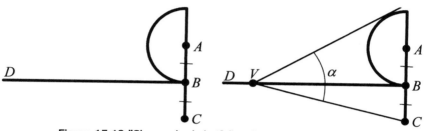

Figure 15.12 "Shoemaker's knife" or "tomahawk" trisector

In order to give a correct statement of what is impossible, it is necessary first to define a **compass and unmarked straightedge sequence** to be a finite sequence of points, lines, and circles that starts with two distinct given points and such that (1) Each of the other points in the sequence is the intersection of lines or circles that occur before it in the sequence. (2) Each circle has its center and one point on the boundary occurring before it in the sequence. (3) Each line contains two distinct points that occur before it in the sequence.

Theorem 15.4. *It is impossible to trisect a 60° angle with a compass and unmarked straightedge sequence.*

For a proof of this theorem and related discussions about various compass and straightedge constructions, see [**TX**: Martin] or [**TX**: Hartshorne], Section 28.

Chapter 16

INVERSIONS IN CIRCLES

Q: How does a geometer capture a lion in the desert?

A: Build a circular cage in the desert, enter it, and lock it. Now perform an inversion with respect to the cage. Then you are outside and the lion is locked in the cage.

— A mathematical joke from before 1938

We now study inversions in a circle (which is the analogue of reflection in a line) and its applications. Though inversion in a circle can be defined on spheres and hyperbolic planes, it seems to have no significant applications on these surfaces. Therefore, in this chapter we will only consider the case of the Euclidean plane.

To study Chapter 16, the only results needed in Chapters 10–15 are

PROBLEM 13.4: The **AAA similarity** and **SAS similarity** criteria for triangles on the plane.

PROBLEM 15.1b: *On a plane, if two lines through a point P intersect a circle at points A, A' (possibly coincident) and B, B' (possibly coincident), then*

$$|PA| \times |PA'| = |PB| \times |PB'| .$$

If you are willing to assume these criteria for similar triangles, then you can work through Chapter 16 without Chapters 10–15.

EARLY HISTORY OF INVERSIONS

Apollonius of Perga (c. 250–175 B.C.) was famous in his time for work on astronomy (Navigation/Stargazing Strand), before his now

217

well-known work on conic sections. Unfortunately, Apollonius' original work on astronomy and most of his mathematical work (except for *Conics*, [**AT**: Apollonius]) has been lost and we only know about it from a commentary by Pappus of Alexandria (290–350 A.D.). According to Pappus, Apollonius investigated one particular family of circles and straight lines. Apollonius defined the curve:

> $c_k(A, B)$ is the locus of points P such that $PA = k \times PB$, where A and B are two points in the Euclidean plane, and k is a positive constant.

Now this curve is a straight line if $k = 1$ and a circle otherwise and is usually called an ***Apollonian circle***. Apollonius proved (see Problem **16.1b**) that a circle c (with center C and radius r) belongs to the family $\{c_k(A,B)\}$ if and only if $BC \times AC = r^2$ and A and B are on the same ray from C. In modern terms, we use "$BC \times AC = r^2$ and A and B are on the same ray from C" as the definition of A and B being inversions of each other with respect to the circle c. There is indirect evidence (see [**HI**: Calinger], page 181) that Apollonius used inversion in a circle to solve an astronomical problems concerning celestial orbits. The theory of inversions was apparently not carried on in a systematic way until the 19th century, when the theory was developed purely geometrically from Euclid's Book III, but this (as far as we know) was not done in ancient times. We suggest that this was because Euclid's *Elements* and Apollonius' circles were parts of different historical strands. In Problem **16.4** we will explore a problem of Apollonius that uses inversions for its solutions.

PROBLEM 16.1   INVERSIONS IN CIRCLES

DEFINITIONS. An ***inversion with respect to a circle*** Γ is a transformation from the extended plane (the plane with ∞, the "point at infinity," added) to itself that takes C, the center of the circle, to ∞ and vice versa, and that takes a point at a distance s from the center to the point on the same ray (from the center) that is at a distance of r^2/s from the center, where r is the radius of the circle. See Figure 16.1. We call (P, P') an ***inversive pair*** because (as the reader can check) P and P' are taken to each other by the inversion. The circle Γ is called the *circle of inversion*.

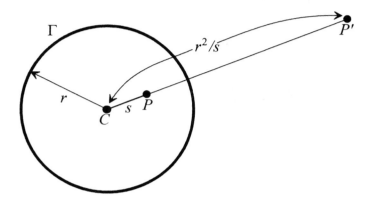

Figure 16.1   Inversion with respect to a circle

Note that an inversion takes the inside of the circle to the outside and vice versa and that the inversion takes any line through the center to itself. Because of this, inversion can be thought of as a reflection in the circle. See also part **c** for the close connection between inversions in the plane and reflections on a sphere.

We strongly suggest that the reader play with inversions by using dynamic geometry software such as *Geometers Sketchpad*®, *Cabri*®, or *Cinderella*®. Sample constructions can be found at the *Experiencing Geometry* Web site: www.math.cornell.edu/~henderson/ExpGeom/. You may construct the image, *P'*, of *P* under the inversion through the circle Γ as follows (see Figure 16.2):

*If P is inside Γ, then draw through P the line perpendicular to the ray CP. Let S and R be the intersections of this line with Γ. Then P' is the intersection of the lines tangent to Γ at S and R. (*To construct the tangents, note that lines tangent to a circle are perpendicular to the radius of the circle.)

*If P is outside Γ, then draw the two tangent lines from P to Γ. Let S and R be the points of tangency on Γ. Then P' is the intersection of the line SR with CP. (*The points *S* and *R* are the intersections of Γ with the circle with diameter *CP*. *Why?*)

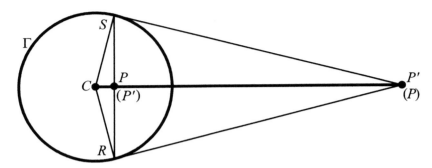

Figure 16.2 Constructing inversive images

a. *Prove that these constructions do construct inversive pairs.*

The purpose of this part is to explore and better understand inversion, but it will not be directly used in the other parts of this problem. When P is inside Γ, you need to prove that C-P-P' are collinear. When P' is outside Γ, you need to prove that SR is perpendicular to CP.

Next we prove

*b. **Apollonius' Theorem.** *Define the curve $c_k(A, B)$ to be the locus of points P such that $PA = k \times PB$, where A and B are two points in the plane and k is a positive constant. A circle c (with center C and radius r) belongs to the family $\{c_k(A, B)\}$ if and only if A and B are an inversive pair with respect to c.*

The following result demonstrates the close connection between inversion through a circle in the plane and reflections through great circles on a sphere. If you have studied Problem **14.4** (or assume it), then you can use this part **c** in your analysis of inversions in **16.2**.

*c. *Let Σ be a sphere tangent at its south pole to the plane Π and let $f\colon \Sigma \to \Pi$ be a stereographic projection from the north pole. If Γ is the circle that is the image under f of the equator and if g is the intrinsic (or extrinsic) reflection of the sphere through its equator (or equatorial plane), then show that the transformation $f \circ g \circ f^{-1}$ is the inversion of the plane with respect to the circle Γ. See Figure 16.3.*

Imagine a sphere tangent to a plane at its south pole, *S*. Now use ***stereographic projection*** to project the sphere from the north pole, *N*, onto the plane; see Problem **14.4**. Stereographic projection was known already to Hipparchus (Greek, second century B.C.). Show that the triangle Δ*SNP* is similar to Δ*RNS*, which is congruent to Δ*QSN*, which is similar to Δ*SP'N*.

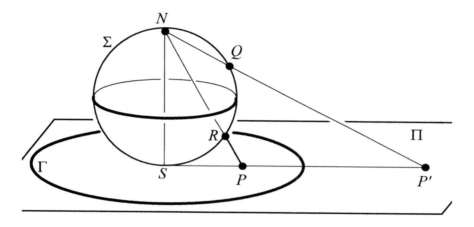

Figure 16.3   Stereographic projection and inversion

PROBLEM **16.2**   INVERSIONS PRESERVE ANGLES AND PRESERVE CIRCLES (AND LINES)

Part **c** provides another route to prove that inversions are conformal (preserves angles, see Problem **14.4**). In Problem **14.4** you showed that *f*, stereographic projection, is conformal. In addition, *g* (being an isometry) is conformal. Thus, the inversion *f* ∘ *g* ∘ *f*⁻¹ is conformal.

    a. *Show that an inversion takes each circle orthogonal to the circle of inversion to itself.* See Figure 16.4.

Two ***circles are orthogonal*** if, at each point of intersection, the angle between the tangent lines is 90°. (Note that, at these points, the radius of one circle is tangent to the other circle.)

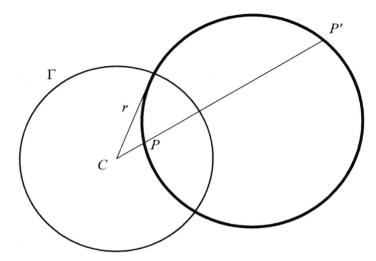

Figure 16.4　A circle orthogonal to Γ inverts to itself

b. *Show that an inversion takes a circle through the center of inversion to a line not through the center, and vice versa. What happens in the special cases when either the circle or the straight line intersects the circle of inversion?*

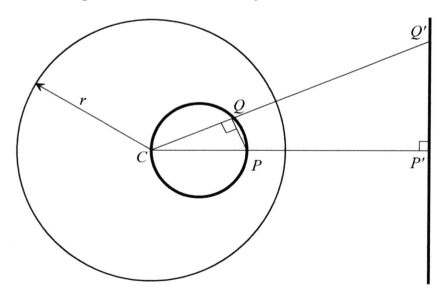

Figure 16.5　Circles through the center invert to lines

Look at Figure 16.5 (where *CP* is a diameter of the circle) and prove that $\triangle CPQ$ and $\triangle CQ'P'$ are similar triangles. Note that the line is parallel to the line tangent to the circle at *C*.

   c. *An inversion takes circles not through the center of inversion to circles not through the center.* Note: The (circumference of a) circle inverts to another circle but the centers of these circles are on the same ray from *C* though **not** an inversive pair.

Look at Figure 16.6, where *PQ* is a diameter of the circle. If *P*, *Q*, *X* invert to *P'*, *Q'*, *X'*, then show that

$$\angle P'\, X'\, Q' = \angle P\, X\, Q = \text{right angle}$$

by looking for similar triangles. Thus, argue that as *X* varies around the circle with diameter *PQ*, then *X'* varies around the circle with diameter *Q'P'*.

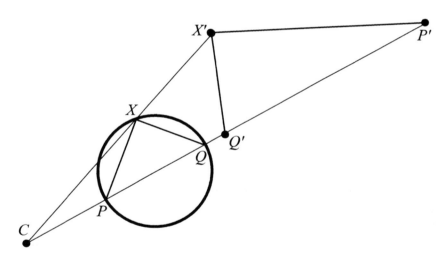

Figure 16.6   Circles invert to circles

   d. *Inversions are conformal.*

Look at two lines that intersect and form an angle at *P*. Look at the images of these lines.

    Inversions were used in the 19th century to solve a long-standing engineering problem (from the Motion/Machines Strand) that is the subject of Problem **16.3**. Other applications (that started with Apollonius) are discussed in Problem **16.4**. More history and further expansions of the notion of inversion are contained in the last section.

Problem **16.3** Using Inversions to Draw Straight Lines

At the beginning of Chapter 1, there is a brief history of attempts to find linkages that would draw a straight line. In this problem we explore the mathematics behind this linkage.

    a. *Show that for the linkage in Figure* 16.7 *the points P and Q are the inversions of each other through the circle of inversion with center at C and radius* $r = \sqrt{s^2 - d^2}$.

Draw the circle with center R and radius d and note that C, P, Q are collinear.

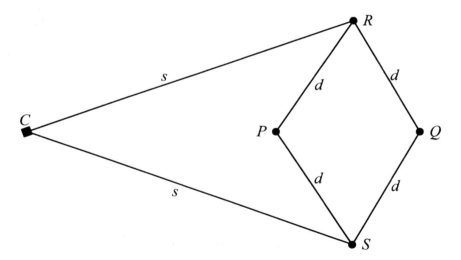

    ■— = Fixed point with hinge
    ●— = Variable point with hinge

Figure 16.7   A linkage for constructing an inversion

b. *Show that the point Q in the linkage in Figure* 16.8 *always traces a straight line.*

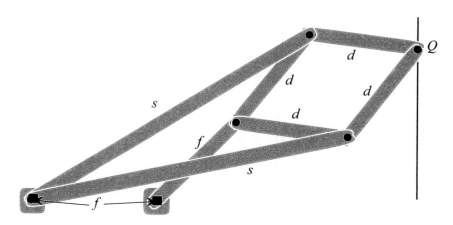

Figure 16.8   Linkage for drawing a straight line

If we modify the Peaucellier-Lipkin linkage by changing the distance between the anchor points, then

c. *The point Q in the linkage in Figure* 16.9 *always traces the arc of a circle. Why? Show that the radius of the circle is expressed by $r^2 f / (g^2 - f^2)$, where r is as in part* **a.**

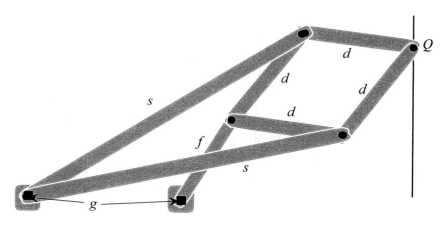

Figure 16.9   Peaucellier-Lipkin linkage modified to draw the arc of a circle

*Problem 16.4   Apollonius' Problem

In Book IV of the *Elements*, Euclid shows how to construct the circle that passes through three given (non-collinear) points and also how to construct a circle tangent to three given straight lines (not passing the same point). Apollonius of Perga (c. 250–175 B.C.) generalized this to

> **Apollonius' Problem:** *Given three objects, each of which may be a point, a line, or a circle, construct a circle that passes through each of the given points and is tangent to the given lines and circles.*

Solutions to this problem are discussed in Apollonius' *On Tangencies* (*De Tactionibus*). Unfortunately, Apollonius' work has not survived, but it has been "reconstructed" from both Arabic and Greek commentaries especially through the description of its contents from Pappus of Alexandria (ca. A.D. 300). In Book 7 of the *Mathematical Collection*, Pappus described the contents of various works by Apollonius. Pappus presents a list of problems in Apollonius' lost work, and on the basis of this information the work has been reconstructed at least four times.

Francois Viete (1540–1603) restored Apollonius *De tactionibus* and published it under the title *Apollonius Gallus* in 1600. Frans van Schooten, in a 1657 reconstruction, showed that Apollonius' problem can be solved by the algebraic methods of Descartes' *Geometrie* (1637). Joachim Jungius and Woldeck Weland (1622–1641), in a reconstruction titled *Apollonius Saxonicus*, used a purely geometrical method that they called "metagoge," that is, the reduction of the general case of a problem to a special case of the same problem or a simpler problem (such as in **16.4c** below). Another reconstruction was completed by Robert Simson (1749). The authors of these reconstructions were doing mathematics, not the history of mathematics, as can be inferred from the fact that the "reconstructions" differ from each other and sometimes deal with generalizations of the problems that had actually been treated by Apollonius.

In addition to the attempted reconstructions, there were many and varied solutions of the Apollonius problem produced by later mathematicians, including Isaac Newton (1643–1727), A. van Roomen (1561–1615), J. Casey (1820–1881), R. Descartes, P. Fermat, Princess Elizabeth (1596–1662), L. Euler (1707–1783), N. Fuss (1755–1826), L. N. M. Carnot (1753–1823), J. D. Gergonne (1771–1859), C. F. Gauss (1777–1855), J. V. Poncelet (1788–1867), A. L. Cauchy (1789–1857), and Eduard Study (1862–1930). You should recognize some of these names. Poncelet and Cauchy solved Apollonius' problem while first year students at the École Polytechnique.

In 1679 P. Fermat formulated and solved an extension of Apollonius problem to 3-space: Construct a sphere tangent to four given spheres. This is called Fermat's problem by some later authors. A large number of mathematicians were discussing this problem in 19th century. Some further generalizations of Apollonius' problem are discussed in H. S. M. Coxeter, "The Problem of Apollonius," *American Mathematical Monthly* 75 (1968), pp. 5–15.

In 2003, R. H. Lewis and S. Bridgett, in a paper entitled "Conic tangency equations and Apollonius problems in biochemistry and pharmacology" (*Mathematics and Computers in Simulation*, vol. 61; Jan. 2003, pp. 101–114), discuss current applications of Apollonius' problems. The applications involve bonding interactions in human bodies between protein molecules and hormone, drug, and other molecules.

Apollonius' problem can be discussed in 10 possible cases (letting P = point, L = line, C = circle):

PPP, PPL, PPC, PLL, PLC, PCC, LLL, LLC, LCC, CCC.

a. *Solve the case* **PPP**. *Show that the solution implies that the perpendicular bisectors of the sides of a triangle all pass through the same point.* We call this point the ***circumcenter*** of the triangle. See Figure 16.10.

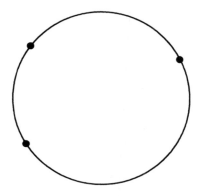

Figure 16.10 **PPP**

Hint: If the three points are on a line, then that line (as a circle with infinite radius) is the solution. Otherwise the three points determine a triangle.

b. *Solve the case* **LLL**. *Show that the solution implies that the angle bisectors of a triangle pass all through the same point.* This point is called the ***incenter*** of the triangle.

Hint: If the three lines intersect in the same point, then there is no solution. If the three lines form a triangle, then there are four solutions. See Figure 16.11. What happens in other cases?

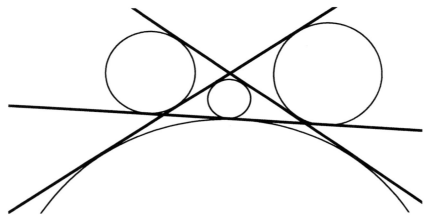

Figure 16.11 **LLL**

c. *Solve the cases* **PPL** *and* **PPC**.

Outline of solution: If both points are on the given line or circle, the only solution is the given line or circle. If point *A* is not on the given line (or circle), then we can take *A* as the center of an inversion. Denote the other point *B* and the given line or circle Γ. Under the inversion, Γ goes to another circle Γ' (never a line, *Why?*) and *B* goes to another point *B'*. Let *l* be a line through *B'* that is tangent to Γ'. Inverting this tangent *l* back to the original picture, we will have a solution. (*Why?*) If *B* is on Γ, then there exists one solution; if neither A or B are on Γ, then there are two solutions.

d. *Solve the cases* **PLL**, **PLC**, *and* **PCC**.

Hint: Choose the given point as the center of an inversion. After the inversion the solution will be a line (in order to contain the inversive image of the point). See Figure 16.12. Depending on the original location of the given point, we get the following subcases:

1. The point is on neither of the lines (line and circle, circles). Then those circles (line and circle, lines) in inversion will go to two circles and the problem reduces to constructing a tangent to two given circles. In this subcase there are either 0, 1, 2, 3, or 4 solutions possible. (*Why?*)
2. The point is the point of tangency of circles (or circle and line).
3. The point is the point of intersection of circles (or circle and line).
4. The point is on one circle (or line) but not on the other.

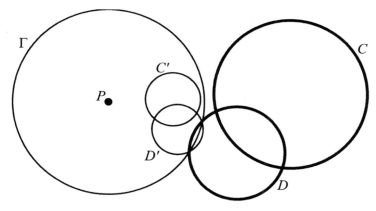

Figure 16.12 Case **PCC**: invert through Γ and then find tangent to *C*′and *D*′

e. Application of PCC. *There is a story that in World War I troops used* **PCC** *to pinpoint the location of large enemy guns. The three separate observation points synchronize their clocks to the second and then note the exact second that they hear the gun's sound. How could these timings be used to pinpoint the location of the gun?*

Hint: The speed of sound is approximately 340 meters/second (it varies ±5% in somewhat predictable ways with temperature and atmospheric conditions). Draw a picture of the instant in time when the sound reaches the first observation point.

f. *Solve cases* **LLC, LCC,** *and* **CCC.**

Outline of solution: Each of these three cases can be reduced to either CCC or PCC by using an appropriate inversion. (*Do you see how?*) There are many subcases depending on how the circles and lines relate to each other: inside, outside, intersecting. However, the overall strategy is to reduce these cases to **PLC**, **PLL**, or **PCC** in part **d**. We illustrate this with the subcase of **LCC**, where all circles and lines are disjoint and the two circles are on the same side of the line and exterior to each other. See Figure 16.13.

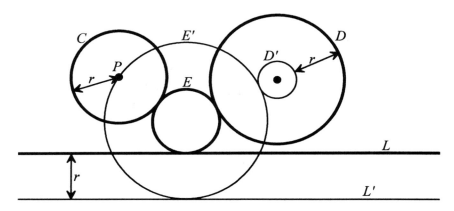

Figure 16.13 Reducing **LCC** to **PLC**

Let r be the radius of the smaller circle, C. Shrink the other circle, D, to the circle, D', with the same center but with radius reduced by r. Construct another line parallel to L, on the side opposite to C and D, and at a distance of r. Now apply **PLC** to the point P, the circle D', and the line L', to get a circle E' that is tangent to P, D', and L'. The required solution to the original **LCC** is the circle E with the same center as E' but with radius decreased by r. See Figure 16.13.

EXPANSIONS OF THE NOTION OF INVERSIONS

There was a rebirth of interest in inversions in the 19th century. Jakob Steiner (1796–1863) was among the first to start extensively using the technique of inversions in circles to solve geometric problems. Steiner had no early schooling and did not learn to read or write until he was age 14. Against the wishes of his parents, at age 18 he went to the Pestalozzi School at Yverdon, Switzerland, where his extraordinary geometric intuition was discovered. By age 28 he was making many geometric discoveries using inversions. At age 38 he occupied the chair of geometry established for him at the University of Berlin, a post he held until his death.

Steiner defined an ***inversive transformation*** to be any transformation that is the composition of inversions and initiated ***inversive geometry***, which is the study of properties of the extended plane that are preserved by inversive transformations. It follows from Problem **16.2** that inversive transformations preserve angles and take circles and lines to circles and lines.

Jean Victor Poncelet (1788–1867) showed that an inversion is a ***birational transformation***, that is, a one-to-one transformation of the extended plane such that both the transformation and its inverse are of the form

$$x' = f(x,y), \quad y' = g(x,y), \quad \text{where } f \text{ and } g \text{ are rational functions.}$$

There were many other European mathematicians in the 19th century who studied inversions and inversive geometry. Applications of inversions in physics were used by Lord Kelvin (Sir William Thomson) (1824–1907) in 1845, and also by Joseph Liouville (1809–1882) in 1847, who called inversions *the transformations by reciprocal radii.*

In 1854 Luigi Cremona (1830–1903) made a systematic study of birational transformations that carry the entire extended plane onto itself. These transformations are now often called ***Cremona transformations***. Since inversions take the entire extended plane to itself, they are Cremona transformations. These were subsequently studied by Max Noether (1844–1921), who proved that a plane Cremona transformation (and thus inversions) could be constructed by a sequence of quadratic and linear transformations.

In 1855, August Ferdinand Möbius (1790–1868) undertook a systematic study of circular transformations (conformal transformations that map points on a circle to points on a circle) by purely geometrical means. He defined what are now called ***Möbius transformations***, which are often studied today in courses on complex analysis:

> A ***Möbius transformation*** *is any transformation of the extended complex plane onto itself of the form*

$$M(z) = \frac{az+b}{cz+d}, \text{ where } a, b, c, d \text{ are complex numbers and } ad - bc \neq 0.$$

The following properties of Möbius transformations are proved in Sections 5.3 and 5.4 of [**TX**: Brannan]:

◆ *Every Möbius transformation is an inversive transformation* (but no inversion can be a Möbius transformation because Möbius transformations preserve orientation and inversions do not).

◆ Möbius *transformations form a subgroup of the group of inversive transformation.*

◆ *Every inversion F can be written in the form* $F(z) = M(\bar{z})$, *where M is a Möbius transformation.*

◆ *Given any two sets of three points, z_1, z_2, z_3, and w_1, w_2, w_3, there is a unique Möbius transformation that maps z_1 to w_1, z_2 to w_2, and z_3 to w_3.*

Möbius geometry is also connected to Laguerre geometry initiated by Edmond Laguerre (1834–1868) and Minkowskian geometry initiated by Hermann Minkowski (1864–1909), which is the geometry that Einstein used in the Theory of Relativity and Space/Time. In 1900, Edward Kasner (1878–1955) was apparently the first to study inversive geometry in accordance with Klein's Erlanger Program (see Chapter 11, page 153). Research on transformations that preserve circles continued into at least the middle of the 20th century. Continuing to today, inversions are used in the two Poincaré models of hyperbolic geometry (see Problems **17.2** through **17.5**).

For more discussion of inversive and related geometries, see [**TX**: Brannan], Chapter 5, and [**HI**: Kline], Section 39.3. Kline also makes connections to algebraic geometry. See also [**HM**: Marchisotto] for related history in the 20th century.

Chapter 17

PROJECTIONS (MODELS) OF HYPERBOLIC PLANES

> In recent times the mathematical public has begun to occupy itself with some new concepts which seem to be destined, in the case they prevail, to profoundly change the entire order of classical geometry.
>
> — E. Beltrami (1868), when he developed
> the projective disk model (Problem **17.5**)

In this chapter we will study projections of a hyperbolic plane onto the plane and use these "models" to prove some results about the geometry of hyperbolic planes. In the case of hyperbolic planes it is customary to call these "***models***" instead of "projections" because it was thought that there were no surfaces that were hyperbolic planes. As in the case of spherical projections, any projections (models) of the hyperbolic plane must distort some geometric properties; and with models it is more difficult to gain the intrinsic and intuitive experiences that are possible with the hyperbolic surfaces discussed in Chapter 5. Nevertheless, these models do give the most analytically accurate picture of hyperbolic planes and allow for more accurate and precise constructions and proofs. We take as our starting point the geodesic rectangular coordinates presented in Problem **5.2**. In order to connect these coordinates to the study of the models, we will need the results on circles from Chapter 14 and an analytic sophistication that is not necessary in other chapters in this book. However, no technical results from analysis are needed. The reader may bypass most of the analytic technicalities (which occur in Problems **17.1** and **17.2**) if the reader is willing to assume the results of Problem **17.2**, which make the connections between an annular hyperbolic plane and the upper-half-plane model and prove which curves in the upper half-plane correspond to geodesics in the annular hyperbolic plane. The basic properties of geodesics and constructions in the upper-half-plane model (and therefore in annular hyperbolic planes) are investigated in Problem **17.3**. We continue our work on the area of triangles by investigating in Problem **17.4** ideal and 2/3-ideal triangles. Other popular models of hyperbolic planes are contained in **17.5** (Poincaré disk model) and **17.6** (projective disk model).

DISTORTION OF COORDINATE SYSTEMS

The reader should review the description of the annular hyperbolic plane in Chapter 5 and the discussion in **5.2**. In Problem **5.2**, we defined geodesic rectangular coordinates on the annular hyperbolic plane as the map $x: R^2 \to H^2$ defined as indicated in Figure 17.1.

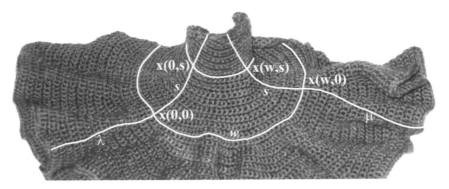

Figure 17.1   Geodesic rectangular coordinates on annular hyperbolic plane

We showed in Problem **5.2** that the coordinate map x is one-to-one and onto from the whole of R^2 onto the whole of the annular hyperbolic plane. Horizontal lines map onto the annular strips and vertical lines map onto radial geodesics. Then we showed the following:

> **5.2b.** *Let λ and μ be two radial geodesics on a hyperbolic plane with radius ρ. If the distance between λ and μ along the base curve is w, then the distance between them at a distance s from the base curve is*
>
> $$w \exp(-s/\rho).$$

Thus, the coordinate chart x preserves (does not distort) distances along the (vertical) second coordinate curves but at $x(a, b)$ the distances along the first coordinate curve are distorted by the factor of $\exp(-b/\rho)$ when compared to the distances in $\mathbf{R}^2$. To be more precise,

DEFINITION. Let $y: A \rightarrow B$ be a map, and let $t \mapsto \lambda(t)$ be a curve in A. Then the **distortion** of y along λ at the point $p = \lambda(0)$ is defined as

$$\lim_{t \to 0} \frac{\text{the arc length along } y(\lambda) \text{ from } y(\lambda(t)) \text{ to } y(\lambda(0))}{\text{the arc length along } \lambda \text{ from } \lambda(t) \text{ to } \lambda(0)}.$$

In the case of the above coordinate curves, λ is the path in $\mathbf{R}^2$, $t \mapsto (a + t, b)$ or $t \mapsto (a, b + t)$, and the distortions of x along the coordinate curves are

$$\lim_{t \to 0} \frac{\text{the arc length from } x(a + t, b) \text{ to } x(a, b)}{|(a + t, b) - (a, b)|} =$$

$$= \frac{t \exp(-b/\rho)}{t} = \exp(-b/\rho)$$

and

$$\lim_{t \to 0} \frac{\text{the arc length from } x(a, b + t) \text{ to } x(a, b)}{|(a, b + t) - (a, b)|} = \frac{t}{t} = 1.$$

We seek a change of coordinates that will distort distances equally in both coordinate directions. The reason for seeking this change (as we will see below) is that if distances are distorted the same in both coordinate directions, then the chart will preserve angles. (Remember, we call such a chart *conformal*.)

We cannot hope to have no distortion in both coordinate directions (if there were no distortion, then the chart would be an isometry), so we try to make the distortion in the second coordinate direction the same as the distortion in the first coordinate direction. After a little experimentation we find that the desired change is

$$z(x,y) = x(x, \rho \ln(y/\rho)),$$

with the domain of z being the upper half-plane

$$\mathbf{R}^{2+} \equiv \{ (x,y) \in \mathbf{R}^2 \mid y > 0 \},$$

where x is the geodesic rectangular coordinates defined above. This is usually called the **upper-half-plane model** of the hyperbolic plane. The upper-half-plane model is a convenient way to study the hyperbolic plane — think of it as a map of the hyperbolic plane in the same way that we use planar maps of the spherical surface of the earth.

*PROBLEM 17.1   A CONFORMAL COORDINATE SYSTEM

Show that the distortion of z along both coordinate curves

$$x \rightarrow z(x, b) \quad and \quad y \rightarrow z(a, y)$$

at the point z(a, b) is ρ/b.

It may be best to first try this for $\rho = 1$. For the first coordinate direction, use the result of Problem **5.2b**. For the second coordinate direction, use the fact that the second coordinate curves in geodesic rectangular coordinates are parametrized by arc length. Use first-semester calculus where necessary.

> **Lemma 17.1.** *If the distortion of z at the point p = (a, b) is the same [say Δ(p)] along each coordinate curve, then at (a, b) the distortion of z has the same value along any other curve λ(t) = z(x(t), y(t)) that passes through p; and z preserves angles at p (that is, z is conformal).*

Proof. Suppose that $\lambda(0) = (x(0), y(0)) = (a, b) = p$. Assuming that the annular hyperbolic plane can be locally isometrically (that is, preserving distances and angles) embedded in 3-space (see Problem **5.3**), the distortion of z along λ at p is

$$\lim_{t \to 0} \frac{\text{the arc length along } z(\lambda) \text{ from } z(\lambda(t)) \text{ to } z(\lambda(0))}{\text{the arc length along } \lambda \text{ from } \lambda(t) \text{ to } \lambda(0)} =$$

$$= \lim_{t \to 0} \frac{\frac{1}{|t|}[\text{the arc length along } z(\lambda) \text{ from } z(\lambda(t)) \text{ to } z(\lambda(0))]}{\frac{1}{|t|}[\text{the arc length along } \lambda \text{ from } \lambda(t) \text{ to } \lambda(0)]} =$$

$$= \frac{\text{speed of } z(\lambda(t)) \text{ at } t = 0}{\text{speed of } \lambda(t) \text{ at } t = 0} = \frac{\left|\frac{d}{dt}z(\lambda(t))\right|_{t=0}}{\left|\frac{d}{dt}\lambda(t)\right|_{t=0}} = \frac{|(z \circ \lambda)'(0)|}{|\lambda'(0)|}.$$

Therefore, along the first coordinate curve $t \mapsto (a + t, b)$, the distortion is

$$\frac{\left|\frac{d}{dt}z(a+t,b)\right|_{t=0}}{\left|\frac{d}{dt}(a+t,b)\right|_{t=0}} = \text{the norm of the partial derivative, } |z_1(p)|.$$

Similarly, the distortion along the second coordinate curve is $|z_2(p)|$. The velocity vector of the curve $z(\lambda(t)) = z(x(t),y(t))$ at p is

$$\tfrac{d}{dt}z(x(t),y(t))_{t=0} = z_1(p)\tfrac{d}{dt}x(t)_{t=0} + z_2(p)\tfrac{d}{dt}y(t)_{t=0}.$$

Thus, the velocity vector, $(z \circ \lambda)'$ at $t = 0$ is a linear combination of the partial derivative vectors, $z_1(p)$ and $z_2(p)$ — note that these vectors are orthogonal. Therefore, the velocity vectors of curves through $p = \lambda(0)$ all lie in the same plane called the tangent plane at $z(p)$. Also note that the velocity vector, $(z \circ \lambda)'$, depends only on the velocity vector, $\lambda'(0)$, and not on the curve λ. Thus, z induces a linear map (called the differential dz) that takes vectors at $p = \lambda(0)$ to vectors in the tangent plane at $z(p)$. This differential is a similarity that multiples all length by $\Delta(p)$ and thus preserves angles. The distortion of z along λ is also $\Delta(p)$.

DEFINITION.   In the above situation we call $\Delta(p)$ the ***distortion of the map z at the point p*** and denote it **dist(z)(p)**.

*PROBLEM 17.2   UPPER HALF-PLANE IS MODEL OF ANNULAR HYPERBOLIC PLANE

We were able to prove in Problem **5.1** that there are reflections about the radial geodesics but only assumed (based on our physical experience with physical models) the existence of other geodesics and reflections through them. To assist us in looking at transformations of the annular hyperbolic space (with radius ρ), we use the upper-half-plane model. If f is a transformation taking the upper half plane $\boldsymbol{R}^{2+}$ to itself, then we have

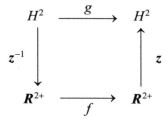

We see that $g = z \circ f \circ z^{-1}$ is a transformation from the annular hyperbolic plane to itself. We call g the transformation of H^2 that ***corresponds*** to f.

We will call *f* an ***isometry of the upper-half-plane model*** if the corresponding *g* is an isometry of the annular hyperbolic plane. To show that *g* is an isometry, you must show that the transformation $g = z \circ f \circ z^{-1}$ preserves distances. Remember that distance along a curve is equal to the integral of the speed along the curve. Thus, it is enough to check that the distortion of *g* at each point is equal to 1. Before we do this we must first show that

> **a.** *The distortion of an inversion i_C with respect to a circle Γ at a point P, which is a distance s from the center C of Γ, is equal to r^2/s^2, where r is the radius of the circle.* See Figure 17.2.

Hint: Because the inversion is conformal, the distortion is the same in all directions. Thus check the distortion along the ray from *C* through *P*. The distance along this ray of an arbitrary point can be parametrized by $t \mapsto ts$. Use the definition of distortion given in Problem **17.1**.

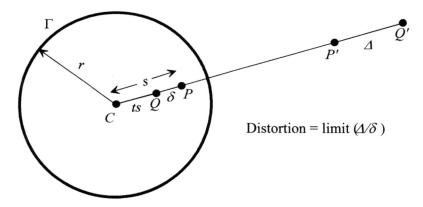

Figure 17.2   Distortion of an inversion

> **b.** *Let f be the inversion in a circle whose center is on the x-axis. Show that f takes $\mathbf{R}^{2+}$ to itself and that $g = z \circ f \circ z^{-1}$ has distortion 1 at every point and is thus an isometry.*

OUTLINE OF A PROOF:

1. Note that each of the maps z, z^{-1}, *f* are conformal and have at each point a distortion that is the same for all curves at that point. If

dist(k)(p) denotes the distortion of the function k at the point p, then argue that

$$\text{dist}(g)(p) = \text{dist}(z^{-1})(p) \times \text{dist}(f)(z^{-1}(p)) \times \text{dist}(z)(f(z^{-1}(p))).$$

1. If $z(a, b) = p$, then show (using **5.1c**) that dist(z^{-1})(p) = b/ρ, where ρ is the radius of the annuli.

2. Show (using part **a**) that dist(f)($z^{-1}(p)$) = r^2/s^2, where r is the radius of the circle C that defines f and s is the distance from the center of C to (a, b).

3. Then show that $\text{dist}(z)(f(z^{-1}(p))) = \dfrac{\rho}{b\frac{r^2}{s^2}}$.

We call these inversions (or the corresponding transformations in the annular hyperbolic plane) **hyperbolic reflections**. We also call reflections through vertical half-lines (corresponding to radial geodesics) hyperbolic reflections.

Now you can prove that

 c. *If γ is a semicircle in the upper half-plane with center on the x-axis or a straight half-line in the upper half-plane perpendicular to x-axis, then $z(\gamma)$ is a geodesic in the annular hyperbolic plane.*

Because of this, we say that such γ are **geodesics in the upper-half-plane model**. Because the compositions of two isometries is an isometry, we see immediately that any composition of inversions in semicircles (whose centers are on the x-axis) is an isometry in the upper-half-plane model (that is, the corresponding transformation in the annular hyperbolic plane is an isometry).

PROBLEM **17.3**   PROPERTIES OF HYPERBOLIC GEODESICS

 a. *Any similarity (dilations) of the upper half-plane corresponds to an isometry of an annular hyperbolic. Such similarities must have their centers on the x-axis. (Why?)*

Look at the composition of inversions in two concentric semicircles.

b. *If γ is a semicircle in the upper half-plane with center on the x-axis, then there is an inversion (in another semicircle) that takes γ to a vertical line that is tangent to γ.*

Hint: An inversion takes any circle through the center of the inversion to a straight line (see Problem **16.2**).

Each of the following three parts is concerned with finding a geodesic. Each problem should be looked at in both the annular hyperbolic plane and in the upper-half-plane model. In a crocheted annular hyperbolic plane we can construct geodesics by folding much the same way we can on a piece of (planar) paper. Geodesics in the upper-half-plane model can be constructed using properties of circles and inversions (see Problem **16.2**). You will also find part **b** very useful.

c. *Given two points A and B in a hyperbolic plane, there is a unique geodesic joining A to B; and there is an isometry that takes this geodesic to a radial geodesic (or vertical line in the upper-half-plane model).*

In the upper-half-plane model, construct a circle with center on the x-axis that passes through A and B. Then use part **b**.

We use AB to denote the unique geodesic segment joining A to B.

d. *Given a geodesic segment AB with endpoints points A and B in a hyperbolic plane, there is a unique geodesic that is a perpendicular bisector of AB.*

Use appropriate folding in annular hyperbolic plane. In the upper-half-plane model, make use of the properties of a reflection through a perpendicular bisector.

e. *Given an angle ∠ABC in a hyperbolic plane, there is a unique geodesic that bisects the angle.*

In the upper-half-plane model, again use the properties of a reflection through the bisector of an angle.

f. *Any two geodesics on a hyperbolic plane either intersect, are asymptotic, or have a common perpendicular.*

Look at two geodesics in the upper-half-plane model that do not intersect in the upper half-plane nor on the bounding x-axis.

PROBLEM **17.4**   HYPERBOLIC IDEAL TRIANGLES

In Problem **7.2** we investigated the area of triangles in a hyperbolic plane. In the process we looked at ideal triangles and 2/3-ideal triangles. We can look more analytically at the ideal triangles. It is impossible to picture the whole of an ideal triangle in an annular hyperbolic plane, but it is easy to picture ideal triangles in the upper-half-plane model. In the upper-half-plane model an *ideal triangle* is a triangle with all three vertices either on the x-axis or at infinity. See Figure 17.3.

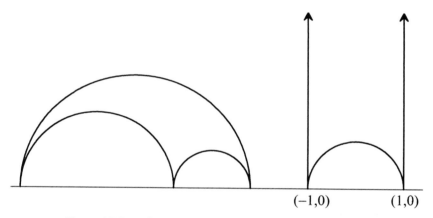

Figure 17.3   Ideal triangles in the upper-half-plane model

At first glance it appears that there must be many different ideal triangles. However,

    a. *Prove that all ideal triangles on the same hyperbolic plane are congruent.*

Review your work on Problem **16.2**. Perform an inversion that takes one of the vertices to infinity and the two sides from that vertex to vertical lines. Then apply a similarity to the upper half-plane, taking it to the standard ideal triangle with vertices $(-1,0)$, $(0,1)$, and ∞.

    b. *Show that the area of an ideal triangle is $\pi\rho^2$.* (Remember this ρ is the radius of the annuli.)

Hint: Because the distortion $\mathrm{dist}(z)(a, b)$ is ρ/b, the desired area is

$$\int_{-1}^{1} \int_{\sqrt{1-x^2}}^{\infty} \left(\frac{\rho}{y}\right)^2 dy\, dx.$$

We now picture in Figure 17.4 **2/3-*ideal triangles*** in the upper-half-plane model.

    c. *Prove that all 2/3-ideal triangles with angle θ are congruent and have area $(\pi - \theta)\rho^2$.*

Show, using Problem **16.2**, that all 2/3-ideal triangles with angle θ are congruent to the standard one at the right of Figure 17.4 and show that the area is the double integral

$$\int_{-1}^{\cos\theta} \int_{\sqrt{1-x^2}}^{\infty} \left(\frac{\rho}{y}\right)^2 dy\, dx.$$

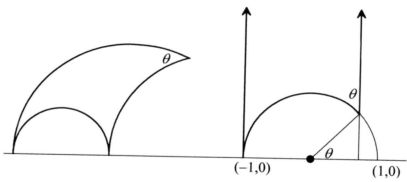

Figure 17.4   2/3-ideal triangles in the upper-half-plane model

PROBLEM 17.5   POINCARÉ DISK MODEL

You showed in Problem **17.1c** that the coordinate map x from a hyperbolic plane to the upper half-plane preserves angles (is conformal); this we called the *upper-half-plane model*. Now we will study other models of the hyperbolic plane.

    Let $z\colon \mathbf{R}^{2+} \to H^2$ be the coordinate map defined in Problem **17.2** that defines the upper-half-plane model. We will now transform the upper-half-plane model to a disk model that was first discussed by Poincaré in 1882.

    a. *Show that any inversion through a circle whose center is in the lower half-plane (that is, y < 0) will transform the upper half-plane onto an open (without its boundary) disk. Show that the hyperbolic geodesics in the upper half-plane are transformed by this inversion into circular arcs (or line segments) perpendicular to the boundary of the disk.*

Review the material on inversions discussed in Problem **16.2**.

    b. *If w: $D^2 \to R^{2+}$ is the inverse of a map from the upper half plane to a (open) disk from part **a**, then show that the composition*

$$z \circ w : D^2 \to H^2$$

is conformal. We call this the (***Poincaré***) ***disk model***, after Henri Poincaré (1854–1912, French).

Review the material on inversions in **16.2** and on the upper-half-plane model in **17.2**.

    c. *Show that any inversion through a circular arc (or line segments) perpendicular to the boundary of D^2 takes D^2 to itself. Show that these inversions correspond to isometries in the (annular) hyperbolic plane.* Thus, we call these circular arcs (or line segments) ***hyperbolic geodesics*** and call the inversions ***hyperbolic reflections*** in the Poincaré disk model.

Review Problem **17.2**.

    See Figure 17.5 for a drawing of geodesics and a triangle in the Poincaré disk and the projective disk model (Problem **17.6**).

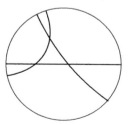
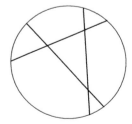

In Poincaré disk model   In projective disk model

Figure 17.5   Geodesics and a triangle

PROBLEM 17.6   PROJECTIVE DISK MODEL

Let D^2 be the disk model of a hyperbolic plane and assume its radius is 2. Then place a sphere of radius 1 tangent to the disk at its center. Call this point of tangency the south pole S. See Figure 17.6.

Now let s be the stereographic projection from the sphere to the plane containing D^2, and note that s(equator) is the boundary of D^2, and thus s takes the Southern Hemisphere onto D^2. Now let h be the orthogonal projection of the southern hemisphere onto the disk, B^2, of radius 1.

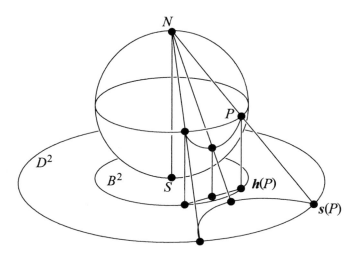

Figure 17.6   Obtaining the projective disk model from the Poincaré disk model

Show that the mapping $h \circ s^{-1}$ takes D^2 to B^2 and takes each circle (or diameter) of D^2 to a (straight) cord of B^2. Thus

$$h \circ s^{-1} \circ (z \circ w)^{-1}$$

is a map from the hyperbolic plane to B^2, which takes geodesics to straight line segments (cords) in B^2.

We call this the **projective disk model**, but in the literature it is also called the **Beltrami/Klein model** or the **Klein model**, named after Eugenio Beltrami (1835–1900, Italian), who described the model in 1868, and Felix Klein (1849–1925, German), who fully developed it in 1871. See Figure 17.5 for a drawing of geodesics and a triangle in the projective disk model.

Chapter 18

GEOMETRIC 2-MANIFOLDS

> The concept "two-dimensional manifold" or "surface" will not be associated with points in three-dimensional space; rather it will be a much more general abstract idea.
>
> — Hermann Weyl (1913)

There is clearly a large variety of different surfaces around in our experiential world. The study of the geometry of general surfaces is the subject of differential geometry. In this chapter we will study *geometric 2-manifolds*, that is, a connected space that locally is isometric to either the (Euclidean) plane, a sphere, or a hyperbolic plane. The surface of a cylinder (no top or bottom and indefinitely long) and a cone (with the cone point removed) are examples of geometric 2-manifolds. We study these because their geometry is simpler and closely related to the geometry we have been studying of the plane, spheres, and hyperbolic planes. In addition, the study of these surfaces will lead us to the study of the possible global shapes of our physical universe.

There are no prerequisites for this chapter from after Chapter 7, but some of the ideas may be difficult the first time around. Problems **18.1**, **18.3**, and **18.6** are the minimum that is needed from this chapter before you study Chapter 24 (3-Manifolds — Shape of Space); the other problems can be skipped.

We use the term "manifold" here instead of "surface" because we usually think of surfaces as sitting extrinsically in 3-space. Here we want to study only the intrinsic geometry; and thus any particular extrinsic embedding does not matter. Moreover, we will study some geometric 2-manifolds (for example, the flat torus) that cannot be (isometrically) embedded in 3-space. We ask what is the intrinsic geometric experience on geometric 2-manifolds of a 2-dimensional bug. How will the bug view geodesics (intrinsically straight lines) and triangles? How can a bug on a geometric 2-manifold discover the global shape of its universe? These questions will help us as we think about how we as human beings can think about our 3-dimensional physical universe, where *we* are the (3-dimensional) bugs.

This chapter will only be an introduction to these ideas. For a geometric introduction to differential geometry, see [**DG**: Henderson]. For more details about geometric 2-manifolds, see [**DG**: Weeks] and Chapter 1 of [**DG**: Thurston]. For the classification of (triangulated) 2-manifolds, see the recent [**TP**: Francis & Weeks], which contains an accessible proof due to John H. Conway.

In Problem **4.1**, we have already studied two examples of geometric 2-manifolds — cylinders and cones (without the cone point). Because these surfaces are locally isometric to the Euclidean plane, these types of geometric manifolds are called *flat* (or *Euclidean*) *2-manifolds*. It would be good at this point for you to review what you know from Chapter 4 about cylinder and cones.

PROBLEM 18.1   FLAT TORUS AND FLAT KLEIN BOTTLE

FLAT TORUS

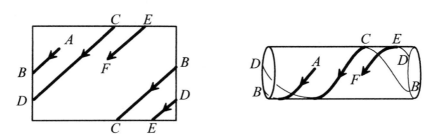

Figure 18.1   1-sheeted covering of cylinder and flat torus

Another example of a flat (Euclidean) 2-manifold is provided by a video game that was popular a while ago. A blip on the video screen representing a ball travels in a straight line until it hits an edge of the screen. Then the blip reappears traveling parallel to its original direction from a point at the same position on the opposite edge. Is this a representation of some surface? If so, what surface? First, imagine rolling the screen into a tube where the top and bottom edges are glued (Figure 18.1). This is a representation of the screen as a 1-sheeted covering of the cylinder. A blip on the screen that goes off the top edge and reappears on the bottom is the lift of a point on the cylinder that travels around the cylinder, crossing the line that corresponds to the joining of the top and bottom of the screen.

Now let us further imagine that the cylinder can be stretched and bent so that we can glue the two ends to make a torus. Now the screen represents a 1-sheeted covering of the torus. If the blip goes off on one side and comes back on the other at the same height, this represents the lift of a point moving around the torus and crossing the circle that corresponds to the place where the two ends of the cylinder are joined. The possible motions of a point on the torus are represented by the motions on the video screen!

You can't make a model in 3-space of a flat torus from a flat piece of paper without distorting it. Such a torus is called a *flat torus*. It is best not to call this a "surface," because there is no way to realize it isometrically in 3-space and it is not the surface of anything. But the question of whether or not you can make an isometric model in 3-space is not important — the point is that the gluings in Figure 18.1 intrinsically define a flat 2-manifold.

If you distort the cylinder in Figure 18.1 in 3-space, you can get the torus pictured in Figure 18.2. This is not a geometric 2-manifold because the original flat (Euclidean) geometry has been distorted and it is also not exactly either spherical or hyperbolic.

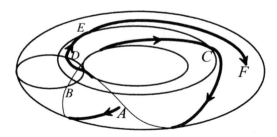

Figure 18.2   Non-flat torus

a. *Show that the flat torus is locally isometric to the plane and thus, is a geometric 2-manifold, in particular, a flat (Euclidean) 2-manifold.*

Note that each point on the interior of an edge of the screen is the lift of a point that has another lift on the opposite edge. Thus, a lift of a neighborhood of that is in two pieces (one near each of the two opposite edges). What happens at the four corners of the computer screen (which are lifts of the same point)?

The torus in Figure 18.2 and the flat torus are related in that there is a continuous one-to-one mapping from either to the other. We say that they are **homeomorphic**, or **topologically equivalent**. We can further express this situation by saying that the torus in Figure 18.2 and the flat torus are both **topological tori**.

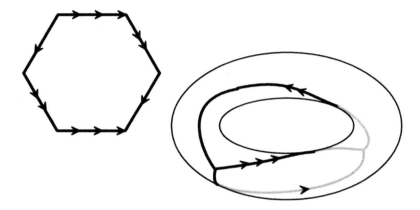

Figure 18.3   Flat torus from a hexagon

There is another representation of the flat torus based on a hexagon. Start with a regular hexagon in the plane and glue opposite sides as indicated in Figure 18.3.

b. *Show that gluing the edges of the hexagon as in Figure* 18.3 *forms a flat 2-manifold (called the **hexagonal torus**).*

In order for a flat 2-manifold to be formed from a regular polygon, the interior angle of the polygon must be an integral factor of 360°. (*Do you see why?*) Thus it is not possible to use other regular polygons to create flat 2-manifolds; see Problem **11.7**. However, we can see from **11.7** that it may be possible for spherical and/or hyperbolic 2-manifolds to be created from regular polygons; see Problems **18.3** and **18.4**.

*Flat Klein Bottle

Now we describe a related geometric 2-manifold, traditionally called a *flat Klein bottle*, named after Felix Klein (1849–1925, German). Imagine the same video screen, again with a traveling blip representing a ball that travels in a straight line until it hits an edge of the screen. When it hits the top edge then the blip proceeds exactly the same as for the flat torus (traveling parallel to its original direction from a point at the same position on the opposite edge). However, when the blip hits a vertical edge of the screen it reappears on the opposite edge but in the *diametrically opposite* position and travels in a direction with slope that is the negative of the original slope. (See Figure 18.4.) As before, imagine rolling the screen into a tube where the top and bottom edges are joined. This is again a representation of the screen as a 1-sheeted covering of the cylinder.

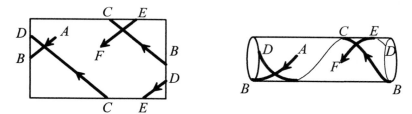

Figure 18.4   1-sheeted covering of a flat Klein bottle

A blip on the screen that goes off the top edge and reappears on the bottom is the lift of a point on the cylinder that travels around the cylinder crossing the line that corresponds to the joining of the top and bottom of the screen. Now the screen represents a 1-sheeted covering of the Klein bottle. The possible motions of a point on the Klein bottle are represented by the motions on the video screen!

Now, let us further imagine that the cylinder can be stretched and bent so that we can join the two ends to make a topological Klein bottle; see Figure 18.5. The figure in Figure 18.5 is not a geometric 2-manifold, because most points do not have neighborhoods that are isomorphic to either the plane, a sphere, or a hyperbolic plane.

You also can't make an isometric model in 3-space of a flat Klein bottle without distorting it and having self-intersections. But the gluings in Figure 18.4 define intrinsically a flat 2-manifold.

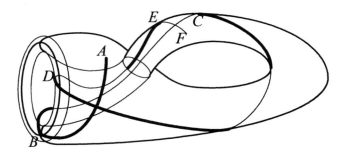

Figure 18.5   Topological Klein bottle

*c. *Show that the flat Klein bottle is locally isometric to the plane and thus is a geometric 2-manifold, in particular, a flat (Euclidean) 2-manifold.*

Note that the four corners of the video screen are lifts of the same point and that a neighborhood of this point has 360° — that is, 90° from each of the four corners.

It can be shown that

THEOREM 18.1. *Flat tori and flat Klein bottles are the only flat (Euclidean) 2-manifolds that are finite and geodesically complete (every geodesic can be extended indefinitely).* See the last section in Chapter 4, Relations to Differential Geometry.

For a detailed discussion, see [**DG:** Thurston], pages 25–28. For a more elementary discussion, see [**DG:** Weeks], Chapters 4 and 11.

Note that a finite cylinder is not geodesically complete; and if it is extended indefinitely, then it is geodesically complete but not finite. A cone with the cone point is not a flat manifold at the cone point; with the cone point removed the cone is not geodesically complete.

Note that we get a flat torus for each size rectangle in the plane. These flat tori are different geometrically because there are different distances around the tori. However, topologically they are all the same as (homeomorphic to) the surface of a doughnut.

Note that if you move a right-hand glove (which we stylize by ℵ) around the flat torus, it will always stay right handed; however, if you move it around the flat Klein bottle horizontally, it will become left handed. See Figure 18.6. We describe these phenomena by saying that the flat torus is *orientable* and a Klein bottle is ***non-orientable***.

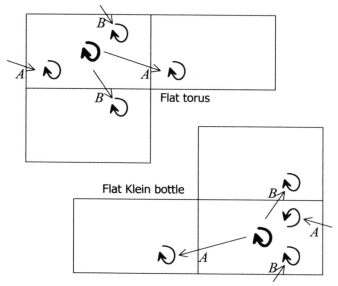

Figure 18.6   Orientable and non-orientable

*PROBLEM **18.2**   UNIVERSAL COVERING OF FLAT 2-MANIFOLDS

Problem **4.2** contains an introduction to some of the ideas in this problem and thus we urge the reader to look at **4.2** before proceeding with this chapter.

a. *On a flat torus or flat Klein bottle, how do we determine the different geodesics connecting two points? How many are there? How can we justify our conjectures?*

Look at straight lines in the universal coverings introduced below.

b. *Show that some geodesics on the flat torus or flat Klein bottle are closed curves (in the sense that they come back and continue along themselves like great circles), though possibly self-intersecting. How can you find them?*

Look in the universal coverings introduced below.

    c.   *Show that there are geodesics on the flat torus and flat Klein bottle that never come back and continue along themselves.*

Look at the slopes of the geodesics found in part **b**.

    The geodesics found in part **c** can be shown to come arbitrarily close to *every* point on the manifold. Such curves are said to be ***dense*** in the manifold.

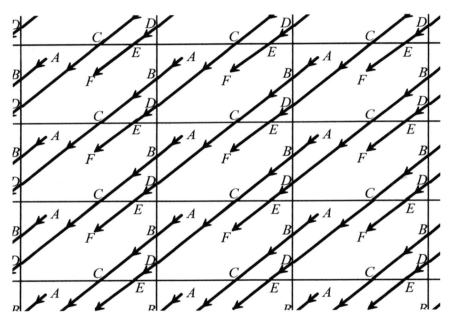

Figure 18.7   Universal covering of a flat torus

SUGGESTIONS

We suggest that you use coverings just as you did for cones and cylinders. The difference is that in this case the sheets of the coverings extend in two directions. See Figure 18.7 for a covering of the flat torus. If this covering is continued indefinitely in all directions, then the whole plane covers the flat torus with each point in the torus having infinitely many lifts. When a covering is the whole of either the (Euclidean) plane, a sphere, or a hyperbolic plane, it is called ***the universal covering***. See

Figure 18.8 for a universal covering of a flat Klein bottle. These coverings are called "universal" because there are no coverings of the plane, spheres, or hyperbolic planes that have more than one sheet; see the next section for a discussion of this for a sphere.

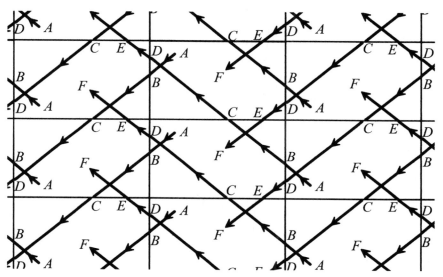

Figure 18.8   Universal covering of the flat Klein bottle

PROBLEM **18.3**   SPHERICAL **2-MANIFOLDS**

Start by considering another version of the video screen as depicted in Figure 18.9.

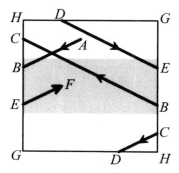

Figure 18.9   This is *not* a geometric 2-manifold

Imagine the same video screen, again with a traveling blip representing a ball that travels in a straight line until it hits an edge of the screen. When the blip reaches *any* edge of the screen, it reappears on the opposite edge but in the *diametrically opposite* position and travels in a direction with slope that is the negative of the original slope. (See Figure 18.9.)

a. *Show that the situation in Figure 18.9 does not represent a geometric 2-manifold because the corners represent two cone points with cone angles 180°.*

Cut around the corners marked *G* and tape together the edges as indicated.

If you restrict yourself to the shaded strip in Figure 18.9 and identify the left and right edges as indicated, then you obtain a ***Möbius strip***, named after August Möbius (German, 1790–1868). You may have seen this surface before — if not, you should be sure to construct one from a (preferably long) strip of paper. To have some fun with a Möbius strip, take a pencil and draw a line in a middle of your Möbius strip parallel to the edges. What happens? Cut your strip in half, following this line in a middle. What do you get? Make another Möbius strip and cut it starting about 1/3 from the edge of a strip and cut parallel to edges. What happens now? The Möbius strip fails to be a (flat) geometric manifold only because it has an edge (note that there is only one edge!); but it is an example of what is called a "flat geometric manifold with boundary."

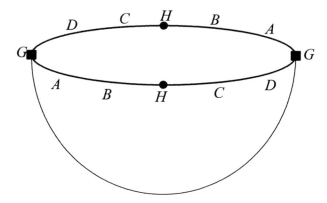

Figure 18.10   Gluings on a hemisphere producing a projective plane

The gluings (on all four edges) indicated in Figure 18.9 fail to produce a geometric 2-manifold because the interior angles are only 90°. It seems that we might get a geometric manifold if the interior angles were 180°. Thus, we need a quadrilateral with equal opposite sides and 180° interior angles. There is no such quadrilateral in the plane; however, on the sphere there *is* such a quadrilateral! See Figure 18.10.

What we have in Figure 18.10 is a hemisphere with each point of the bounding equator being glued to its antipode. In this way we get what is called the (real) projective plane, often denoted **RP**^{2}.

b. *Show that a projective plane is a spherical 2-manifold.*

Examine the neighborhood of a point of the projective plane that comes from the bounding equator. Note that there is a "strip" connecting *B-H-C* to itself along the hemisphere that is very similar to the Möbius strip except that it is a *spherical* manifold with boundary.

c. *What are the geodesics on a projective plane?*

It is clear that the geodesics come from half great circles in the hemisphere, but what happens as one of these half great circles is crossing the equator that is glued?

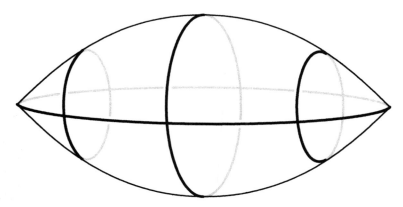

Figure 18.11 A spherical cone

If we cut out and remove a lune from a sphere and then join together the two edges of the lune, we have what can reasonably be called a spherical cone. See Figure 18.11. Note that at least some of these spherical cones have a shape similar to an American football.

By pasting together several (equal radius) spheres with the same lune removed, you can get multiple-sheeted branched coverings of a spherical cone. A spherical cone with the two cone points removed is a finite spherical 2-manifold, but (as with ordinary cones) it is not geodesically complete.

  d. *Show that the spherical cones as described above are spherical 2-manifolds if you remove the two cone points.*

  e. *Identify the geodesics on a spherical cone with cone angle* 180° *(that is, you remove from the sphere a lune with angle* 360° − 180° = 180°*). What happens with other cone angles?*

Look at the great circles in the sphere minus the lune before its edges are joined to produce the spherical cone.

It can be shown that

THEOREM 18.3. *Spheres and projective planes are the only spherical 2-manifolds that are finite and geodesically complete (every geodesic can be extended indefinitely).*

For a detailed discussion, see [**DG**: Thurston], pages 25–28. For a more elementary discussion, see [**DG**: Weeks], Chapters 4 and 11.

*COVERINGS OF A SPHERE

There is no way to construct a covering of a sphere that has more than one sheet unless the covering has some "branch points." A ***branch point*** on a covering is a point such that every neighborhood (no matter how small) surrounding the point contains at least two lifts of some point. In any covering of a cone with more than one sheet, the lift of the cone point is a branch point, as you can see in Figure 18.12.

Notice that the coverings of a cylinder and a flat torus have no branch points. For a sphere the matter is very different — any covering of a sphere will have a branch point. You can see this if you try to construct a cover by slitting two spheres, as depicted in Figure 18.13 and then sticking the two together along the slit. The ends of the slit would become branch points. This topic may be explored further in textbooks on geometric or algebraic topology.

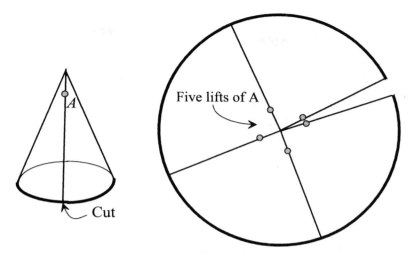

Figure 18.12   Covering space of a cone has branch points

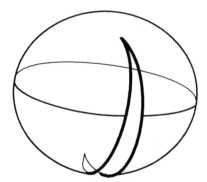

Figure 18.13   Covering space of a sphere has branch points

In fact, any surface that has no (non-branched) coverings and that is bounded and without an edge can be continuously deformed (without tearing) into a round sphere. The surfaces of closed boxes and of footballs are two examples. A torus is bounded and without an edge, but it cannot be deformed into a sphere. A cylinder also cannot be deformed into a sphere, and a cylinder either has an edge or (if we imagine it as extending indefinitely) it is unbounded.

A 3-dimensional analog of this situation arises from a famous, long-unsolved problem called the *Poincaré conjecture.* The analog of a surface is called a *3-dimensional manifold,* a space locally like

Euclidean 3-space (in the same sense that a surface is locally like the plane). The 3-sphere, which we will study in Chapter 22, is a 3-dimensional manifold. Also, in Chapter 24 we note that our physical 3-dimensional universe is a 3-dimensional manifold. Henri Poincaré (1854–1912, French) conjectured that any 3-dimensional manifold that has no (non-branched) coverings and that is bounded and without boundary must be homeomorphic to a 3-dimensional sphere S^3. For the past 80 years, numerous mathematicians have tried to decide whether Poincaré's conjecture is true. On May 24, 2000, the Clay Mathematics Institute announced (www.claymath.org) that it was offering a $1,000,000 prize for a solution of the Poincaré conjecture! In 2002, Grigori Perelman of St. Petersburg, Russia, announced that he had discovered a proof of the Poincaré conjecture. All the details of Perelman's proof have not yet been worked out (as of spring 2004), so it is still too early to tell if he will win the million-dollar prize, but he seems to at least be much closer than anyone else has been. If his ideas work, Perelman will also have settled completely the question of which topological 3-manifolds are geometric 3-manifolds. For discussion of Perelman's ideas and their relationships to the Poincaré conjecture, see articles by J. Milnor (*Notices of the AMS*, Nov. 2003, pp. 1226–1233) and M. T. Anderson (*Notices of the AMS*, Feb. 2004, pp. 184–193). See Chapter 22 and [**EG**: Hilbert] and [**DG**: Weeks] for more discussion of 3-dimensional manifolds and the 3-dimensional sphere.

*PROBLEM 18.4   HYPERBOLIC MANIFOLDS

Is it possible to make a two-holed torus (sometimes called an anchor ring, or the surface of a two-holed donut) into a geometric 2-manifold? See Figure 18.14.

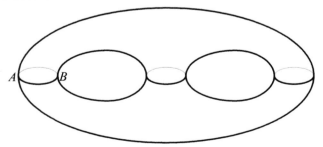

Figure 18.14   Two-holed torus — not a geometric manifold

Note that as it is pictured in Figure 18.14 the two-holed torus is definitely **not** a geometric 2-manifold because the intrinsic geometry is not the same at every point — for example, points A and B. But can we distort the geometry so that the surface is a geometric 2-manifold?

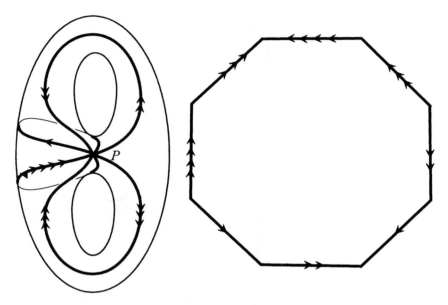

Figure 18.15   Cutting a two-holed torus into an octagon

Imagine cutting the two-holed torus along the four loops emanating from the point P, as indicated in Figure 18.15. You will get a distorted octagon with 45° (= 360°/8) interior angles at each vertex. This distorted octagon is topologically equivalent to a regular planar octagon. Walking around the point P, we find the gluings as indicated in Figure 18.15. Be sure you understand how the gluings on the octagon were determined from the loops on the two-holed torus. If you glue the edges of the regular planar octagon together as indicated in Figure 18.15, you will get a version of the two-holed torus that is locally isometric to the plane except at the (one) vertex. (*Why? What will a neighborhood of the vertex look like?*)

In the plane all octagons have the same interior angle sum. But in the hyperbolic plane, regular octagons have different angles. In fact, we can find such an octagon in the hyperbolic plane with 45° interior angles. (See Figure 18.16.) If we glue the edges of this octagon as indicated, then we will get a hyperbolic 2-manifold. An intermediate step in this process (when every other side is identified) is the so-called "pair of pants," which is on the right-hand side of the front cover.

To see that there is such an octagon, imagine placing a small (regular) octagon on the hyperbolic plane. Because the octagon is small, its interior angles must be very close to the interior angles of an octagon in the (Euclidean) plane. Because the exterior angle of a planar octagon must be 360°/8, the interior angle must be 180° − (360°/8) = 135°. Now let the small octagon grow, keeping it always regular. From Problem **7.2** (remembering that an octagon can be divided into triangles as in Figure 18.16), we conclude that the interior angles of the octagon will decrease in size until, if we let the vertices go to infinity, the angles would decrease to zero. Somewhere between 135° and 0° the interior angles will be the desired 45°.

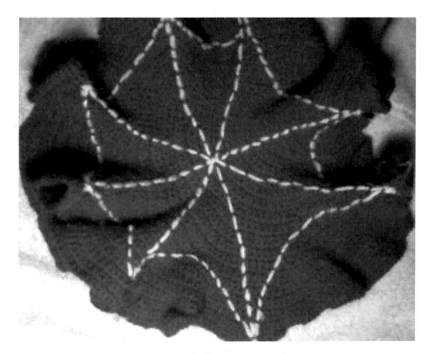

Figure 18.16   Hyperbolic octagon with 45° angles

a. *What is the area of the hyperbolic octagon with 45° interior angles?*

b. *Why is the two-holed torus obtained from a hyperbolic octagon with 45° interior angles a hyperbolic 2-manifold? What is its area?*

It is not easy to determine the geodesics on a hyperbolic 2-manifold. Some discussion about these issues is contained in William Thurston's *Three-Dimensional Geometry and Topology*, Volume 1 [**DG**: Thurston] and will presumably be continued in the forthcoming Volume 2.

There are other ways of making a two-holed torus into a geometric 2-manifold but it is always a *hyperbolic* 2-manifold. However, there are many different hyperbolic structures for a two-holed torus; for example, look in Figure 18.17 for a different way to represent the two-holed torus — this time as the gluing of the boundary of a dodecagon with 90° interior angles.

c. *Follow the steps above to check that Figure 18.17 leads to the representation of a two-holed torus as a dodecagon* (with 90° interior angles) *with gluing on the boundary and thus to a hyperbolic 2-manifold. What is its area?*

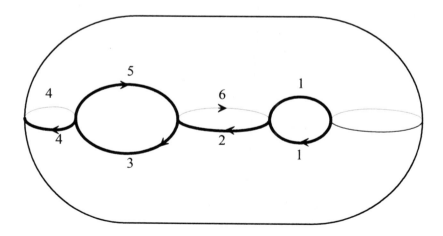

Figure 18.17 Cutting two-holed torus into a dodecagon

You should have found the same area in parts **b** and **c**. In fact, we (you!) will show in the next problem that *any hyperbolic geometric manifold structure on a two-holed torus has the same area.*

All of the above discussion can be extended in straightforward ways to tori with 3, 4, and more holes.

PROBLEM 18.5   AREA AND EULER NUMBER

Now is a good time to go back and review the material on area and holonomy in Problems **7.1–7.5a**. Recall that you showed that the area of any polygon is

$$K \, Area(\Gamma) = [\, \Sigma\beta_i - (n-2)\pi \,],$$

where K is the Gaussian curvature equal to $1/\rho^2$ for a sphere and $-1/\rho^2$ for a hyperbolic plane of radius ρ, $\Sigma\beta_i$ is the sum of the interior angles, and n is the number of edges.

We will now use those results to study the area of geometric manifolds. All of the geometric manifolds that we have described above have cell divisions. A **0-*cell*** is a point we usually call a *vertex*. A **1-*cell*** is a straight-line segment we usually call an *edge* (the edge need not be straight, but in most of our applications it will be). A **2-*cell*** is a polygon that is usually called a *face*. We say that a geometric manifold has a **cell division** if it is divided into cells so that every edge has its boundary consisting of vertices and every face has its boundary divided into edges and vertices and two cells only intersect on their boundaries. We call the cell division a **geodesic cell division** if all the edges are geodesic segments, and thus the faces are polygons. For example, in Figure 18.17 we have a two-holed torus divided into one face, six edges, and three vertices. Note that some edges have only one vertex and thus form a loop (circle). If we make this two-holed torus into a hyperbolic manifold (by using a regular hyperbolic dodecagon with 90° angles), then the cell division is a geodesic cell division.

Suppose we have a geodesic cell division of a geometric manifold M into f faces (2-cells), where the jth face has n_j edges. We can calculate the area of M as follows:

$K \, \text{area}(M) =$

$$= K\{\text{area}(1^{st}\text{ face})\} + K\{\text{area}(2^{nd}\text{ face})\} + \ldots + K\{\text{area}(f^{th}\text{ face})\}$$

$$= \{\Sigma\beta_i \,(1^{st}\text{ face}) - (n_1-2)\pi\} + \ldots + \{\Sigma\beta_i \,(f^{th}\text{ face}) - (n_f-2)\pi\}$$

$$= \{\Sigma\beta_i \,(1^{st}\text{ face}) + \ldots + \Sigma\beta_i \,(f^{th}\text{ face})\} - \{(n_1-2)\pi + \ldots + (n_f-2)\pi\}$$

= {sum of all the angles} $- (n_1 + n_2 + ... + n_f)\pi + (2 + 2 + ... + 2)\pi$.

Fill in the steps to prove the following:

a. *If a geometric manifold M has a geodesic cell division, then*

$$K\,\text{area}(M) = 2\pi(v - e + f),$$

where the cell division has v vertices and e edges and f faces. Check that this agrees with your results in Problems **18.1**, **18.3**, *and* **18.4**.

The quantity $(v - e + f)$ is called the ***Euler number*** (or sometimes the ***Euler characteristic***). This quantity is named after the mathematician Leonhard Euler (1707–1783), who was born and educated in Basel, Switzerland but worked in St. Petersburg, Russia, and Berlin, Germany. It follows directly from part **a** and what we know about area and curvature that

b. *The Euler number of any geodesic cell division of a sphere must be 2. The Euler number of any geodesic cell division of a projective plane must be 1. The Euler number of any geodesic cell division of a flat (Euclidean) 2-manifold must be equal to 0. The Euler number of any geodesic cell division of a hyperbolic 2-manifold must be negative.*

We see from part **b** that in the cases of a sphere, a projective plane, or a flat 2-manifold, the Euler number does not depend on the specific cell division (with geodesic edges). So we can talk about the ***Euler number of the sphere*** (= 2) and the ***Euler number of the torus*** or ***Klein bottle*** (= 0). We also saw above that the two different cell divisions that were given of the two-holed torus have the same area and thus the same Euler number. Can we prove that for every hyperbolic 2-manifold the Euler number (and therefore the area) depends only on the topology and not on the particular cell division? In fact,

THEOREM 18.5a. *The Euler number of any cell division of a 2-manifold depends only on the topology of the manifold and not on the specific cell division. Furthermore, two 2-manifolds*

are homeomorphic if and only if they have the same Euler number and are either both orientable or both non-orientable.

Proofs of this result are somewhat fussy and involve much of the foundational results of topology that were only developed in the 20th century. See Imre Lakatos's *Proofs and Refutations* [**PH**: Lakatos] for an accessible and interesting account of the long and complicated history and philosophy of the Euler number. Lakatos describes an imaginary class discussion about the Euler number in which the tortuous route students take toward a proof mirrors the actual route that mathematicians took. Other proofs are with different additional assumptions. For example, [**DG**: Thurston] (Propositions 1.3.10 and 1.3.12) gives an accessible proof assuming that the reader has some familiarity with vector fields on differentiable manifolds. In Sections 2.4 and 2.5 of [**TP**: Blackett], there is a combinatorial-based proof that assumes the topological 2-manifold has some cell division.

*TRIANGLES ON GEOMETRIC MANIFOLDS

Clearly, if on a flat (Euclidean) 2-manifold a triangle is contained in a region that is isometric to the plane, then the triangle is a planar triangle and has all the properties of a triangle in the plane. The same can be said about triangles in spherical and hyperbolic 2-manifolds. In fact, it can be shown (see any topology text that deals with covering spaces) that

> **THEOREM 18.5b.** *If* Δ *is a triangle in a Euclidean [spherical, hyperbolic] 2-manifold, M, such that* Δ *can be shrunk to a point in the interior of* Δ*, then* Δ *(and its interior) can be lifted to the plane [sphere, hyperbolic plane] that is the universal covering space of M; and thus* Δ *has all the same properties of a triangle in the plane [sphere, hyperbolic plane].*

It is natural to be uncomfortable using covering spaces, but covering spaces are a helpful tool for thinking intrinsically. Some triangles, even though they look strange extrinsically, will look like reasonable triangles for the bug. In Figure 18.18 we give an example of an extrinsically strange triangle that intersects itself but that can be considered a normal triangle from an intrinsic point of view. In fact, it is a planar triangle. Such triangles have all the properties of plane triangles, including SAS and ASA.

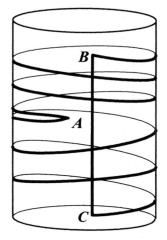

Unroll to a 3-sheeted cover, and ...

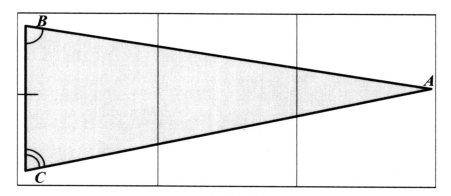

This is a triangle!

Figure 18.18   Think intrinsically

PROBLEM 18.6   CAN THE BUG TELL WHICH MANIFOLD?

Our physical universe is apparently a geometric 3-manifold. In Chapter 24 we will explore ways in which we (human beings) may be able to determine the global shape and size of our physical universe. But first we look at the situation of a 2-dimensional (2-D) bug on a geometric 2-manifold in order to get some help for our 3-dimensional question.

a.   *Suppose a 2-D bug lives on a geometric 2-manifold M and suppose that M is the bug's whole universe. How can the bug determine intrinsically what the local geometry of its universe is?*

b.   *How can the bug in part* **a** *tell what the global shape of its universe is? For example, how can the bug tell (intrinsically) the difference between being on a flat torus and being on a hexagonal torus? Again, for this part you may imagine that the bug can crawl over the whole manifold and leave markers and make measurements.*

c.   *Suppose that the bug in part* **a** *can only travel in a very small region of the manifold (so small that all triangles in the region are indistinguishable from planar triangles), but the bug can see for very long distances. Can the bug still determine which geometric 2-manifold is its universe?*

You must assume that light travels on geodesics in the bug's universe (which is the case of light in our physical 3-dimensional universe). Or forget about light and think about vibrations that travel along the surface. Our 2-D bugs have antennae with which they can receive these vibrations and through them "see" the surface. Imagine that there are bright stars on the 2-manifold that the bug can see; but remember that these stars must be on the 2-manifold — we are taking an intrinsic view of 2-manifolds. You may also assume that there are many stars and they are distributed (roughly) uniformly.

This is different from the situation on the earth, where we have the extrinsic observation of stars (and sun and moon) to help us. The sun was used to measure the radius of the earth by Eratosthenes of Cyrene (Egyptian, 276–194 B.C.) and others. Mariners have used the sun, stars, and eclipses (and clocks) since early times for navigating on the oceans. For an interesting and accessible historical account of the problem of determining one's longitude on the earth, see *Longitude* by Dava Sobel [**CE:** Sobel].

Chapter 19

GEOMETRIC SOLUTIONS OF QUADRATIC AND CUBIC EQUATIONS

> Whoever thinks algebra is a trick in obtaining unknowns has
> thought it in vain. No attention should be paid to the fact that
> algebra and geometry are different in appearance. Algebras
> (jabbre and maqabeleh) are geometric facts which are proved
> by Propositions Five and Six of Book Two of [Euclid's]
> *Elements*.
>
> — Omar Khayyam, a paper [**AT**: Khayyam 1963]

In this chapter we will see how the results from Chapter 13 were used historically to solve equations. Quadratic equations were solved by "completing the square" — a real square. These results in turn lead to conic sections and cube roots and culminate in the beautiful general method from Omar Khayyam that can be used to find all the real roots of cubic equations. Along the way we shall see clearly some of the ancestral forms of our modern Cartesian coordinates and analytic geometry. We will point out several inaccuracies and misconceptions that have crept into the modern historical accounts of these matters. We urge you not to look at this only for its historical interest but also for the meaning it has in our present-day understanding of mathematics. This path is not through a dead museum or petrified forest; it passes through ideas that are very much alive and have something to say to our modern technological, increasingly numerical, world.

PROBLEM 19.1 QUADRATIC EQUATIONS

Finding square roots is the simplest case of solving quadratic equations. If you look in some history of mathematics books (e.g., [**HI**: Joseph] and [**HI**: Eves]), you will find that quadratic equations were solved extensively by the Babylonians, Chinese, Indians, and Greeks. However, the earliest known general discussion of quadratic equations took place between 800 and 1100 A.D. in the Muslim Empire. Best known are Mohammed Ibn Musa al'Khowarizmi (who lived in Baghdad from 780 to 850 and from whose name we get our word "algorithm") and Omar Khayyam (1048–1131), the Persian geometer who is mostly known in the West for his philosophical poetry *The Rubaiyat*). Both wrote books whose titles contain the phrase *Al-jabr w'al mugabalah* (from which we get our word "algebra"), al'Khowarizmi in about 820 and Khayyam in about 1100. (See [**HI**: Katz], Section 7.2.1.) An English translation of both books is available in many libraries, if you can figure out whose name it is catalogued under (see [**AT**: al'Khowarizmi] and [**AT**: Khayyam 1931]). We met Khayyam in Chapter 12.

In these books you find geometric and numerical solutions to quadratic equations and geometric proofs of these solutions. But you will notice quickly that there is not one general quadratic equation as we are used to it:

$$ax^2 + bx + c = 0.$$

Rather, because the use of negative coefficients and negative roots was avoided, these books list six types of quadratic equations (we follow Khayyam's lead and set the coefficient of x^2 equal to 1):

1. $bx = c$, which has root $x = c/b$,

2. $x^2 = bx$, which has root $x = 0$ and $x = b$,

3. $x^2 = c$, which has root $x = \sqrt{c}$,

4. $x^2 + bx = c$, with root $x = \sqrt{(b/2)^2 + c} - b/2$,

5. $x^2 + c = bx$, with roots $x = b/2 \pm \sqrt{(b/2)^2 - c}$, if $c < (b/2)^2$, and

6. $x^2 = bx + c$, with root $x = b/2 + \sqrt{(b/2)^2 + c}$.

Here b and c are always positive numbers or a geometric length (b) and area (c).

a. *Show that these are the only types. Why is $x^2 + bx + c = 0$ not included? Explain why b must be a length but c an area.*

The avoidance of negative numbers was widespread until a few hundred years ago. In the 16th century, European mathematicians called the negative numbers that appeared as roots of equations "numeri fictici" — fictitious numbers (see [**AT:** Cardano, page 11]). In 1759 Francis Masères (1731–1824, English), mathematician and a fellow at Cambridge University and a member of the Royal Society, wrote in his *Dissertation on the Use of the Negative Sign in Algebra,*

> ... [negative roots] serve only, as far as I am able to judge, to puzzle the whole doctrine of equations, and to render obscure and mysterious things that are in their own nature exceeding plain and simple.... It were to be wished therefore that negative roots had never been admitted into algebra or were again discarded from it: for if this were done, there is good reason to imagine, the objections which many leaned and ingenious men now make to algebraic computations, as being obscure and perplexed with almost unintelligible notions, would be thereby removed; it being certain that Algebra, or universal arithmetic, is, in its own nature, a science no less simple, clear, and capable of demonstration, than geometry.

More recently in 1831, Augustus De Morgan (1806–1871, English), the first professor of mathematics at University College, London, and a founder of the London Mathematical Society, wrote in his *On the Study and Difficulties of Mathematics,*

> The imaginary expression $\sqrt{-a}$ and the negative expression $-b$ have this resemblance, that either of them occurring as solution of a problem indicates some inconsistency or absurdity. As far as real meaning is concerned, both are equally imaginary, since $0 - a$ is as inconceivable as $\sqrt{-a}$.

Why did these mathematicians avoid negative numbers and why did they say what they said? To get a feeling for why, think about the meaning of 2×3 as two 3's and 3×2 as three 2's and then try to find a meaning for $3 \times (-2)$ and $-2 \times (+3)$. Also consider the quotation at the beginning of this chapter from Omar Khayyam about algebra and geometry. Some historians have quoted this passage but have left out all the words appearing after "proved." In our opinion, this omission changes the meaning of the passage. Euclid's propositions that are mentioned by Khayyam are the basic ingredients of Euclid's proof of the square root construction and form a basis for the construction of conic sections — see Problem **19.2**, below. Geometric justification when there are negative coefficients is at the least very cumbersome, if not impossible. (If you doubt this, try to modify some of the geometric justifications below.)

> **b.** *Find geometrically the algebraic equations that express all the positive roots of each of the six types. Fill in the details in the following sketch of Khayyam's methods for Types 3–6.*

For the geometric justification of Type 3 and the finding of square roots, Khayyam refers to Euclid's construction of the square root in Proposition II 14, which we discussed in Problem **13.1**.

For Type 4, Khayyam gives as geometric justification the illustration shown in Figure 19.1.

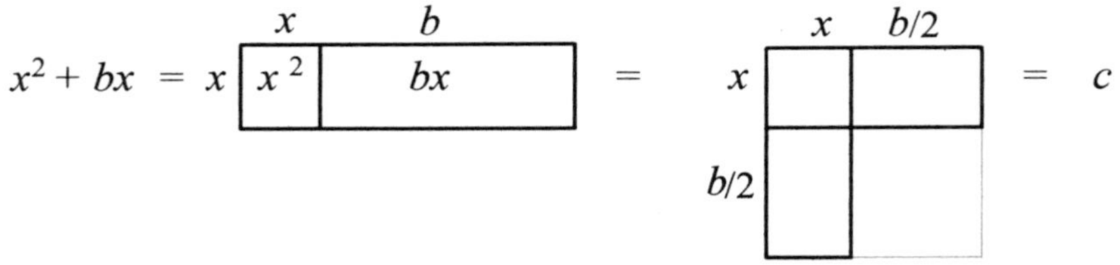

Figure 19.1   Type 4

Thus, by "completing the square" on $x + b/2$, we have

$$(x + b/2)^2 = c + (b/2)^2.$$

Thus we have $x = \ldots$. Note the similarity between this and Baudhayana's construction of the square root (see Chapter 13).

For Type 5, Khayyam first assumes $x < (b/2)$ and draws the equation as Figure 19.2.

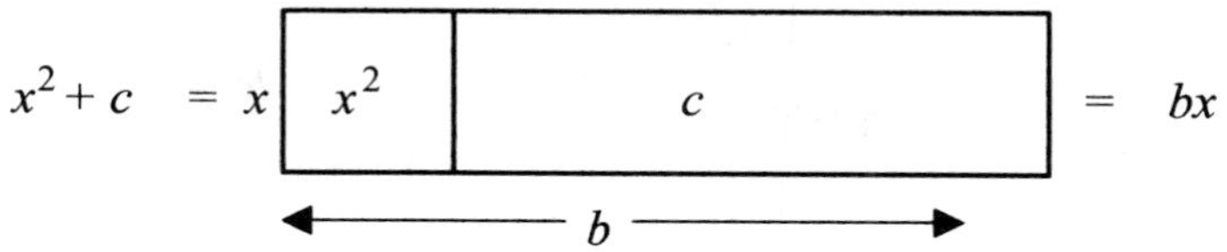

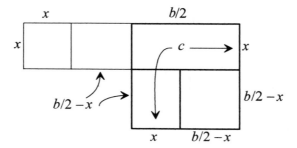

$$x^2 + c \quad = \quad x \; \boxed{\begin{array}{c|c} x^2 & c \end{array}} \quad = \quad bx$$

Figure 19.2 Type 5, $x < (b/2)$

Note (see Figure 19.3) that the square on $b/2$ is $(b/2 - x)^2 + c$.

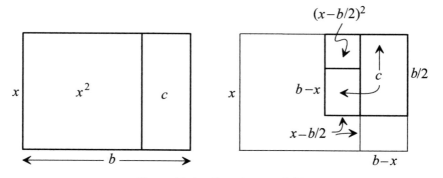

Figure 19.3 Type 5, $x < (b/2)$

This leads to $x = b/2 + \sqrt{(b/2)^2 - c}$. Note that if $c > (b/2)^2$, this geometric solution is impossible. When $x > (b/2)$, Khayyam uses the drawings shown in Figure 19.4.

Figure 19.4 Type 5, $x > (b/2)$

For solutions of Type 6, Khayyam uses the drawing in Figure 19.5.

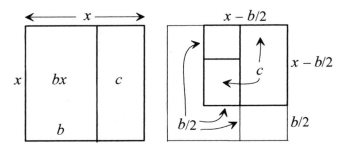

Figure 19.5   Type 6

Do the above solutions find the negative roots? The answer is clearly "no" if you mean, "Did al'Khowarizmi and Khayyam mention negative roots?" But let us not be too hasty. Suppose $-r$ (r positive) is the negative root of $x^2 + bx = c$. Then $(-r)^2 + b(-r) = c$ or $r^2 = br + c$. Thus r is a positive root of $x^2 = bx + c$! The absolute value of the negative root of $x^2 + bx = c$ is the positive root of $x^2 = bx + c$ and vice versa. Also, the absolute values of the negative roots of $x^2 + bx + c = 0$ are the positive roots of $x^2 + c = bx$. So, in this sense, *Yes, the above geometric solutions do find all the real roots of all quadratic equations.* Thus it is misleading to state, as most historical accounts do, that the geometric methods failed to find the negative roots. The users of these methods did not find negative roots because they did not conceive of them. However, the methods can be used directly to find all the positive and negative roots of all quadratics.

    c.   *Use Khayyam's methods to find all roots of the following equations:* $x^2 + 2x = 2$, $x^2 = 2x + 2$, $x^2 + 3x + 1 = 0$.

PROBLEM 19.2   CONIC SECTIONS AND CUBE ROOTS

The Greeks (for example, Archytas of Tarentum, 428–347 B.C., who was a Pythagorean in southern Italy, and Hippocrates of Chios in Asia Minor, 5th century B.C.) noticed that, if $a/c = c/d = d/b$, then $(a/c)^2 = (c/d)(d/b) = (c/b)$ and, thus, $c^3 = a^2 b$. (For more historical discussion, see [**HI**: van der Waerden], page 150, and [**HI**: Katz], Chapter 2.) Now setting $a = 1$, we see that we can find the cube root of b if we can find c and d such that $c^2 = d$ and $d^2 = bc$. If we think of c and d as being variables and b as a constant, then we see these equations as the equations of two parabolas with perpendicular axes and the same vertex. The Greeks also saw it this way, but first they had to develop the concept of a parabola! The first general construction of conic sections was done by Menaechmus (a member of Plato's Academy) in the 4th century B.C.

To the Greeks, and later Khayyam, if *AB* is a line segment, then *the parabola with vertex B and parameter AB* is the curve *P* such that, if *C* is on *P*, then the rectangle *BDCE* (see Figure 19.6) has the property that $(BE)^2 = BD \cdot AB$. Because in Cartesian coordinates the coordinates of *C* are (*BE*, *BD*), this last equation becomes a familiar equation for a parabola.

Points of the parabola may be constructed by using the construction for the square root given in Chapter 13. In particular, *E* is the intersection of the semicircle on *AD* with the line perpendicular to *AB* at *B*. (The construction can also be done by finding *D′* such that *AB = DD′*. The semicircle on *BD′* then intersects *P* at *C*.) We encourage you to try this construction yourself; it is very easy to do if you use a compass and graph paper.

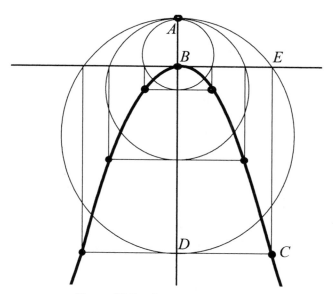

Figure 19.6   Construction of parabola

Now we can find the cube root. Let *b* be a positive number or length and let *AB = b* and construct *C* so that *CB* is perpendicular to *AB* and such that $CB = 1$. See Figure 19.7. Construct a parabola with vertex B and parameter AB and construct another parabola with vertex B and parameter CB. Let E be the intersection of the two parabolas. Draw the rectangle $BGEF$. Then

$$(EF)^2 = BF{\cdot}AB \ \text{ and } \ (GE)^2 = GB{\cdot}CB.$$

But, setting $c = GE = BF$ and $d = GB = EF$, we have

$$d^2 = cb \ \text{ and } \ c^2 = d. \text{ Thus } c^3 = b.$$

If you use a fine graph paper, it is easy to get three-digit accuracy in this construction.

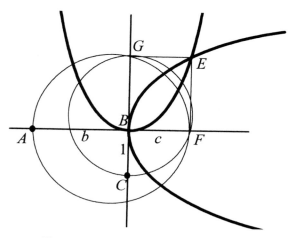

Figure 19.7   Finding cube roots

The Greeks did a thorough study of conic sections and their properties, culminating in Apollonius' (c. 260–170 B.C., Greek) book *Conics*, which appeared around 200 B.C. You can read this book in English translation (see [**AT:** Apollonius]).

    a. *Use the above geometric methods with a fine graph paper to find the cube root of 10.*

To find roots of cubic equations, we shall also need to know *the (rectangular) hyperbola with vertex B and parameter AB*. This is the curve such that, if E is on the curve and $ACED$ is the determined rectangle (see Figure 19.8), then $(EC)^2 = BC \cdot AC$.

The point E can be determined using the construction from Chapter 13. Let F be the bisector of AB. Then the circle with center F and radius FC will intersect at D the line perpendicular to AB at A. From the drawing it is clear how these circles also construct the other branch of the hyperbola (with vertex A).

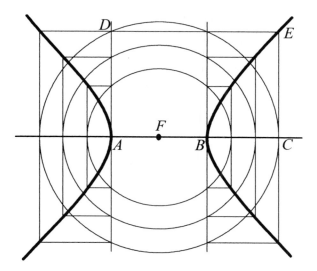

Figure 19.8 Construction of hyperbola

>   **b.**   *Use the above method with graph paper to construct the graph of the hyperbola with parameter 5. What is an algebraic equation that represents this hyperbola?*

Notice how these descriptions and constructions of the parabola and hyperbola look very much as if they were done in Cartesian coordinates. *The ancestral forms of Cartesian coordinates and analytic geometry are evident here.* They are also evident in the solutions of cubic equations in the next section. The ideas of Cartesian coordinates did not appear to Rene Descartes (1596–1650, French) out of nowhere. The underlying concepts were developing in Greek and Muslim mathematics. One of the apparent reasons that full development did not occur until Descartes is that, as we have seen, negative numbers were not accepted. The full use of negative numbers is essential for the realization of Cartesian coordinates. However, even Descartes seems to have avoided negatives as much as possible when he was studying curves — he would start with a curve (constructed by some geometric or mechanical procedure) and then choose axes so that the important parts of the curve had both coordinates positive. However, it is not true (as asserted in some history of mathematics book) that Descartes always used x and y to stand for positive values. For example, in Book II of *Geometrie* [**AT:** Descartes] he describes the construction of a locus generated by a point C and defines on page 60 $y = CB$, where B is a given point, and derives an equation satisfied by y and other variables; and, in the same paragraph on page 63, he continues, "If y is zero or less than nothing in this equation ..."

PROBLEM 19.3   SOLVING CUBIC EQUATIONS GEOMETRICALLY

In his *Al-jabr wa'l muqabalah,* Omar Khayyam also gave geometric solutions to cubic equations. You will see that his methods are sufficient to find geometrically all real (positive or negative) roots of cubic equations; however; in the first chapter Khayyam says (see [**AT:** Khayyam 1931], p. 49),

> When, however, the object of the problem is an absolute number, neither we, nor any of those who are concerned with algebra, have been able to prove this equation — perhaps others who follow us will be able to fill the gap — except when it contains only the three first degrees, namely, the number, the thing and the square.

By "absolute number," Khayyam is referring to what we call algebraic solutions, as opposed to geometric ones. This quotation suggests, contrary to what many historical accounts say, that Khayyam expected that algebraic solutions would be found.

Khayyam found 19 types of cubic equations (when expressed with only positive coefficients). (See [**AT:** Khayyam 1931], p. 51.) Of these 19, 5 reduce to quadratic equations (for example, $x^3 + ax = bx$ reduces to $x^2 + ax = b$). The remaining 14 types Khayyam solved by using conic sections. His methods find all the positive roots of each type, although he failed to mention some of the roots in a few cases, and, of course, he ignored the negative roots. Instead of going through his 14 types, let us look how a simple reduction will reduce them to only four types in addition to types already solved, such as $x^3 = b$. Then we will look at Khayyam's solutions to these four types.

In the cubic $y^3 + py^2 + gy + r = 0$ (where p, g, r, are positive, negative, or zero), set $y = x - (p/3)$. Try it! The resulting equation in x will have the form $x^3 + sx + t = 0$, (where s and t are positive, negative, or zero). If we rearrange this equation so that all the coefficients are positive, we get the following four types that have not been previously solved:

$$(1) \ x^3 + ax = b, \ (2) \ x^3 + b = ax,$$

$$(3) \ x^3 = ax + b, \ \text{and} \ (4) \ x^3 + ax + b = 0,$$

where a and b are positive, in addition to the types previously solved.

    a.   *Show that in order to find all the roots of all cubic equations we need only have a method that finds the roots of Types 1, 2, and 3.*

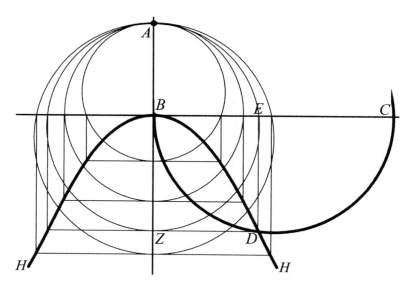

Figure 19.9   Type 1 cubic

KHAYYAM'S SOLUTION FOR TYPE 1: $x^3 + ax = b$

A cube and sides are equal to a number. Let the line AB (see Figure 19.9) be the side of a square equal to the given number of roots [that is, $(AB)^2 = a$, the coefficient]. Construct a solid whose base is equal to the square on AB, equal in volume to the given number [b]. The construction has been shown previously. Let BC be the height of the solid. [That is, $BC \cdot (AB)^2 = b$.] Let BC be perpendicular to AB ... . Construct a parabola whose vertex is the point B ... and parameter AB. Then the position of the conic HBD will be tangent to BC. Describe on BC a semicircle. It necessarily intersects the conic. Let the point of intersection be D; drop from D, whose position is known, two perpendiculars DZ and DE on BZ and BC. Both the position and magnitude of these lines are known.

The root is EB. Khayyam's proof (using a more modern, compact notation) is as follows: From the properties of the parabola (Problem **19.2**) and circle (Problem **15.1**), we have

$$(DZ)^2 = (EB)^2 = BZ \cdot AB \text{ and } (ED)^2 = (BZ)^2 = EC \cdot EB,$$

thus

$$EB \cdot (BZ)^2 = (EB)^2 \cdot EC = BZ \cdot AB \cdot EC$$

and, therefore,

$$AB \cdot EC = EB \cdot BZ,$$

and

$$(EB)^3 = EB \cdot (BZ \cdot AB) = (AB \cdot EC) \cdot AB = (AB)^2 \cdot EC.$$

So

$$(EB)^3 + a(EB) = (AB)^2 \cdot EC + (AB)^2 \cdot (EB) = (AB)^2 \cdot CB = b.$$

Thus EB is a root of $x^3 + ax = b$. Because $x^2 + ax$ increases as x increases, there can be only this one root.

KHAYYAM'S SOLUTIONS FOR TYPES 2 AND 3:
$x^3 + b = ax$ AND $x^3 = ax + b$

Khayyam treated these equations separately; but, by allowing negative horizontal lengths, we can combine his two solutions into one solution of $x^3 \pm b = ax$. Let AB be perpendicular to BC and as before let $(AB)^2 = a$ and $(AB)^2 \cdot BC = b$. Place BC to the left if the sign in front of b is negative (Type 3) and place BC to the right if the sign in front of b is positive (Type 2). Construct a parabola with vertex B and parameter AB.

Construct both branches of the hyperbola with vertices B and C and parameter BC. See Figure 19.10.

Each intersection of the hyperbola and the parabola (except for B) gives a root of the cubic. Suppose they meet at D. Then drop perpendiculars DE and DZ. The root is BE (negative if to the left and positive if to the right). Again, if you use fine graph paper, it is possible to get three-digit accuracy here. We leave it for you, the reader, to provide the proof, which is very similar to Type 1.

b. *Verify that Khayyam's method described above works for Types 2 and 3. Can you see from your verification why the extraneous root given by B appears?*

c. *Use Khayyam's method to find all roots of the cubic*

$$x^3 = 15x + 4.$$

Use fine graph paper and try for three-place accuracy.

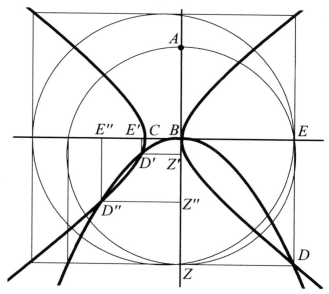

Figure 19.10   Type 2 and type 3 cubics

PROBLEM 19.4   ALGEBRAIC SOLUTION OF CUBICS

A little more history: Most historical accounts assert correctly that Khayyam did not find the negative roots of cubics. However, they are misleading in that they all fail to mention that his methods are fully sufficient to find the negative roots, as we have seen above. This is in contrast to the common assertion (see, for example, [**EM**: Davis & Hersh]) that Girolamo Cardano (1501–1578, Italian) was the first to publish the general solution of cubic equations. In fact, as we shall see, Cardano himself admitted that his methods are insufficient to find the real roots of many cubics.

Cardano published his algebraic solutions in the book *Artis Magnae* (The Great Art) in 1545. For a readable English translation and historical summary, see [**AT**: Cardano]. Cardano used only positive coefficients and thus divided the cubic equations into the same 13 types (excluding $x^3 = c$ and equations reducible to quadratics) used earlier by Khayyam. Cardano also used geometry to prove his solutions for each type. As we did above, we can make a substitution to reduce these to the same types as above:

$$(1)\ \ x^3 + ax = b,\ \ (2)\ \ x^3 + b = ax,$$

$$(3)\ \ x^3 = ax + b,\ \ \text{and}\ \ (4)\ \ x^3 + ax + b = 0.$$

If we allow ourselves the convenience of using negative numbers and lengths, then we can reduce these to one type: $x^3 + ax + b = 0$, where now we allow a and b to be either negative or positive.

The main "trick" that Cardano used was to assume that there is a solution of $x^3 + ax + b = 0$ of the form $x = t^{1/3} + u^{1/3}$. Plugging this into the cubic, we get

$$(t^{1/3} + u^{1/3})^3 + a(t^{1/3} + u^{1/3}) + b = 0.$$

If you expand and simplify this, you get to

$$t + u + b + (3t^{1/3}u^{1/3} + a)(t^{1/3} + u^{1/3}) = 0.$$

(Cardano did this expansion and simplification geometrically by imagining a cube with sides $t^{1/3} + u^{1/3}$.) Thus $x = t^{1/3} + u^{1/3}$ is a root if

$$t + u = -b\ \ \text{and}\ \ t^{1/3}u^{1/3} = -(a/3).$$

Solving, we find that t and u are the roots of the quadratic equation

$$z^2 + bz - (a/3)^3 = 0,$$

which Cardano solved geometrically (and so can you, Problem **19.1**) to get

$$t = -b/2 + \sqrt{(b/2)^2 + (a/3)^3} \quad \text{and} \quad u = -b/2 - \sqrt{(b/2)^2 + (a/3)^3} \ .$$

Thus the cubic has roots

$$x = t^{1/3} + u^{1/3}$$

$$= \{-b/2 + \sqrt{(b/2)^2 + (a/3)^3} \ \}^{1/3} + \{-b/2 - \sqrt{(b/2)^2 + (a/3)^3} \ \}^{1/3}.$$

This is Cardano's cubic formula. But a strange thing happened. Cardano noticed that the cubic $x^3 = 15x + 4$ has a positive real root 4 but, for this equation, $a = -15$ and $b = -4$, and if we put these values into his cubic formula, we get that the roots of $x^3 = 15x + 4$ are

$$x = \{2 + \sqrt{-121} \ \}^{1/3} + \{2 - \sqrt{-121} \ \}^{1/3} \ .$$

But these are the sum of two complex numbers even though you have shown in Problem **19.3** that all three roots are real. How can this expression yield 4?

In Cardano's time there was no theory of complex numbers, and so he reasonably concluded that his method would not work for this equation, even though he did investigate expressions such as $\sqrt{-121}$. Cardano writes ([**AT:** Cardano, p. 103]),

> When the cube of one-third the coefficient of x is greater than the square of one-half the constant of the equation ... then the solution of this can be found by the aliza problem which is discussed in the book of geometrical problems.

It is not clear what book he is referring to, but the "aliza problem" presumably refers to the mathematician known as al'Hazen, Abu Ali al'Hasan ibu al'Haitam (965–1039), who was born in Persia and worked in Egypt and whose works were known in Europe in Cardano's time. Al'Hazen had used intersecting conics to solve specific cubic equations and the problem of describing the image seen in a spherical mirror — this latter problem in some books is called "Alhazen's problem."

In addition, we know today that each complex number has three cube roots and so the formula

$$x = \{2 + \sqrt{-121}\}^{1/3} + \{2 - \sqrt{-121}\}^{1/3}$$

is ambiguous. In fact, some choices for the two cube roots give roots of the cubic and some do not. (Experiment with $x^3 = 15x + 4$.) Faced with Cardano's formula and equations such as $x^3 = 15x + 4$, Cardano and other mathematicians of the time started exploring the possible meanings of these complex numbers and thus started the theory of complex numbers.

a. *Solve the cubic $x^3 = 15x + 4$ using Cardano's formula and your knowledge of complex numbers.*

Remember that on the previous two pages we showed that $x = t^{1/3} + u^{1/3}$ is a root of the equation if $t + u = -b$ and $t^{1/3} u^{1/3} = -(a/3)$.

b. *Solve $x^3 = 15x + 4$ by dividing through by $x - 4$ and then solving the resulting quadratic.*

c. *Compare your answers and methods of solution from Problems* **19.3c**, **19.4a**, *and* **19.4b**.

WHAT DOES THIS ALL POINT TO?

What does the experience of this chapter point to? It points to different things for each of us. We conclude that it is worthwhile to pay attention to the meaning in mathematics. Often in our haste to get to the modern, powerful analytic tools, we ignore and trod upon the meanings and images that are there. Sometimes it is hard even to get a glimpse that some meaning is missing. One way to get this glimpse and find meaning is to listen to and follow questions of "What does it mean?" that come up in ourselves, in our friends, and in our students. We must listen creatively because we and others often do not know how to express precisely what is bothering us.

Another way to find meaning is to read the mathematics of old and keep asking, "Why did they do that?" or "Why didn't they do this?" Why did the early algebraists (up until at least 1600 and much later, we think) insist on geometric proofs? We have suggested some reasons above. Today, we normally pass over geometric proofs in favor of analytic ones based on the 150-year-old notion of Cauchy sequences and the Axiom of Completeness. However, for most students and, we think, most mathematicians, our intuitive understanding of the real numbers is based on the geometric real line. As an example, think about multiplication: What does $a \times b$ mean? Compare the geometric images of $a \times b$ with the multiplication of two infinite, non-repeating, decimal fractions. What is $\sqrt{2} \times \pi$?

There is another reason why a geometric solution may be more meaningful: Sometimes we actually desire a geometric result instead of a numerical one. For example, David and a friend were building a small house using wood. The roof of the house consisted of 12 isosceles triangles that together formed a 12-sided cone (or pyramid). It was necessary for them to determine the angle between two adjacent triangles in the roof so they could appropriately cut the log rafters. David immediately started to calculate the angle using (numerical) trigonometry and algebra. But then he ran into a problem. He had only a slide rule with three-place accuracy for finding square roots and values of trigonometric functions. At one point in the calculation he had to subtract two numbers that differed only in the third place (for example, $5.68 - 5.65$); thus his result had little accuracy. As he started to figure out a different computational procedure that would avoid the subtraction, he suddenly realized *he didn't want a number, he wanted a physical angle*. In fact, a numerical angle would be essentially useless — imagine taking two rough boards and putting them at a given numerical angle apart using only an ordinary protractor! What he needed was the physical angle, full size. So David and his friend constructed the angle on the floor of the house using a rope as a compass. This geometric solution had the following advantages over a numerical solution:

- The geometric solution resulted in the desired physical angle, while the numerical solution resulted in a number.

- The geometric solution was quicker than the numerical solution.

- The geometric solution was immediately understood and trusted by David's friend (and fellow builder), who had almost no mathematical training, while the numerical solution was beyond the friend's understanding because it involved trigonometry (such as the Law of Cosines).

♦   And, because the construction was done full size, the solution automatically had the degree of accuracy appropriate for the application.

Meaning is important in mathematics, and geometry is an important source of that meaning.

Chapter 20

TRIGONOMETRY AND DUALITY

After we have found the equations [The Laws of Cosines and Sines for a Hyperbolic Plane] which represent the dependence of the angles and sides of a triangle; when, finally, we have given general expressions for elements of lines, areas and volumes of solids, all else in the [Hyperbolic] Geometry is a matter of analytics, where calculations must necessarily agree with each other, and we cannot discover anything new that is not included in these first equations from which must be taken all relations of geometric magnitudes, one to another. ... We note however, that these equations become equations of spherical Trigonometry as soon as, instead of the sides a, b, c we put $a\sqrt{-1}$, $b\sqrt{-1}$, $c\sqrt{-1}$ ...

— N. Lobachevsky, quoted in [**HY**: Greenberg]

In this chapter, we will first derive, geometrically, expressions for the circumference of a circle on a sphere, the Law of Cosines on the plane, and its analog on a sphere. Then we will talk about duality on a sphere. On a sphere, duality will enable us to derive other laws that will help our two-dimensional bug to compute sides and/or angles of a triangle given ASA, RLH, SSS, or AAA. Finally, we will look at duality on the plane.

PROBLEM 20.1 CIRCUMFERENCE OF A CIRCLE

a. *Find a simple formula for the circumference of a circle on a sphere in terms of its intrinsic radius and make the formula as intrinsic as possible.*

We suggest that you make an extrinsic drawing (similar to Figure 20.1) of the circle, its intrinsic radius, its extrinsic radius, and the center of the sphere. You may well find it convenient to use trigonometric functions to express your answer. Note that the existence of trigonometric functions for right triangles follows from the properties of similar triangles that were proved in Problem **13.4**.

In Figure 20.1, rotating the segment of length r' (***the extrinsic radius***) through a whole revolution produces the same circumference as rotating r, which is an arc of the great circle as well as ***the intrinsic radius*** of the circle on the sphere.

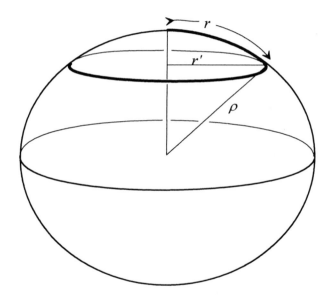

Figure 20.1   Intrinsic radius r

Even though the derivation of the formula this way will be extrinsic, it is possible, in the end, to express the circumference only in terms of intrinsic quantities. Thus, also think of the following problem:

b. *How could our 2-dimensional bug derive this formula?*

By looking at very small circles, the bug could certainly find uses for the trigonometric functions they give rise to. Then the bug could discover that the geodesics are actually (intrinsic) circles, but circles that do not have the same trigonometric properties as very small circles. Then what?

Using the expressions of trigonometric and hyperbolic functions in terms of infinite series, it is proved (in [**HY**: Greenberg], p. 337) that

THEOREM 20.1. *In a hyperbolic plane of radius* 1, *a circle with intrinsic radius r has circumference c equal to*

$$c = 2\pi \sinh(r).$$

c. *Use the theorem to show that on a hyperbolic plane of radius* ρ, *a circle with intrinsic radius r has circumference c equal to*

$$c = 2\pi\rho \sinh(r/\rho).$$

When going from a hyperbolic plane of radius 1 to a hyperbolic plane of radius ρ, all lengths scale by a factor of ρ. *Why?*

The formula in part **c** should look very much like your formula for part **a** (possibly with some algebraic manipulations). This is precisely what Lobachevsky was talking about in the quote at the beginning of this chapter. Check this out with part **d**.

d. *Show that, if you replace ρ by $i\rho$ in the formula of part **c**, then you will get the formula in part **a**.*

Look up the definition of *sinh* (hyperbolic sine) and express it as a Taylor series and then compare with the Taylor series of sine.

PROBLEM **20.2**   LAW OF COSINES

If we know two sides and the included angle of a (small) triangle, then according to SAS the third side is determined. If we know the lengths of the two sides and the measure of the included angle, how can we find the length of the third side? The various formulas that give this length are called the ***Law of Cosines***.

a. *Prove the Law of Cosines for triangles in the plane (see Figure 20.2):*

$$c^2 = a^2 + b^2 - 2ab \cos \theta.$$

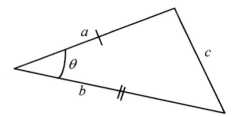

Figure 20.2   Law of Cosines

For a geometric proof of this "law," look at the pictures in Figure 20.3. We first saw the idea for these pictures in the marvelous book [**EG**: Valens]. These pictures show the squares as rigid with hinges at all the points marked ⟋°⟍. Note that in the middle picture θ is greater than $\pi/2$. You must draw a different picture for θ less than $\pi/2$. Prove the Law of Cosines on the plane using the pictures in Figure 20.3, or in any other way you wish. Note the close relationship with the Pythagorean Theorem.

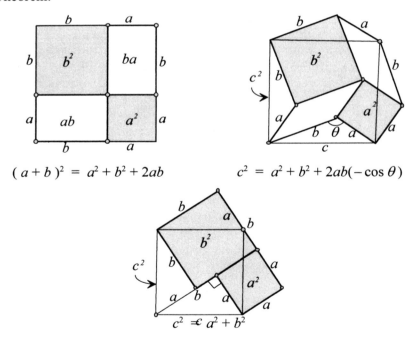

Figure 20.3   Three related geometric proofs

b. *Find the law of cosines for small triangles on a sphere with radius ρ:*

$$\cos c/\rho = \cos a/\rho \cos b/\rho + \sin a/\rho \sin b/\rho \cos \theta.$$

Hint: One approach that will work is to project the triangle by a gnomic projection onto the plane tangent to the sphere at the vertex of the given angle. This projection will preserve the size of the given angle (*Why?*) and, even though it will not preserve the lengths of the sides of the triangle, you can determine what effect it has on these lengths. Now apply the planar Law of Cosines to this projected triangle and turn the algebra crank. It is very helpful to draw a 3-D picture of this projection.

c. *Derive a formula for the distance between two points on the earth in terms of the latitude and longitude at the two points.*

It is sometimes convenient to measure lengths of great circle arcs on the sphere in terms of the radian measure. In particular,

radian measure of the arc = (length of the arc)/ρ,

where ρ is the radius of the sphere. For example, the radian measure of one quarter of a great circle would be $\pi/2$ and the radian measure of half a great circle would be π. In Figure 20.4, the segment a is subtended by the angle α at the pole and by the same angle α at the center of the sphere. The radian measure of a is the radian measure of α. Most other texts and articles either use radian measure for lengths or assume that the radius ρ is equal to 1. We will most of the time keep the ρ explicit so the connection with the radius will be clearly seen.

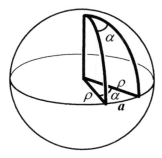

Figure 20.4 Radian measure of lengths

If we measure lengths in radians, then one possible formula for the spherical triangle of radius ρ in part **b** is

$$\cos c = \cos a \cos b + \sin a \sin b \cos \theta.$$

THEOREM 20.2a. *A Law of Cosines for right triangles on a sphere with radius ρ is*

$$\cos c/\rho = \cos a/\rho \cos b/\rho,$$

or, in radian measure of lengths,

$$\cos c = \cos a \cos b,$$

which can be considered as the spherical equivalent of the Pythagorean Theorem.

THEOREM 20.2b. *A Law of Cosines for triangles on a hyperbolic plane with radius ρ is*

$$\cosh c/\rho = \cosh a/\rho \cosh b/\rho + \sinh a/\rho \sinh b/\rho \cos \theta.$$

This theorem is proved in [**HY**: Stahl], p. 125, using analytic techniques, and in [**HY**: Greenberg] using infinite series representations. This is one of the equations that Lobachevsky is talking about in the quote at the beginning of the chapter. It is important to realize that a study of hyperbolic trigonometric functions was started by V. Riccati (1707–1775, Italy) and others, and continued by J. H. Lambert (1728–1777, Germany) in 1761–68 — well before their use in hyperbolic geometry by Lobachevsky. For details on the history of hyperbolic functions, see the informative 2004 paper [**HI**: Barnett].

PROBLEM 20.3 LAW OF SINES

Closely related to the Law of Cosines is the ***Law of Sines***. (Figure 20.5)

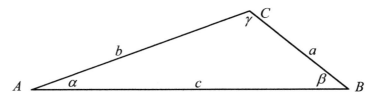

Figure 20.5   Law of Sines

> **a.** *If △ABC is a planar triangle with sides, a, b, c, and corresponding opposite angles, α, β, γ, then a/sin α = b/sin β = c/sin γ.*

The standard proof for the Law of Sines is to drop a perpendicular from the vertex C to the side c and then to express the length of this perpendicular as both (b sin α) and (a sin β). See Figure 20.6. From this the result easily follows. Thus, on the plane the Law of Sines follows from an expression for the sine of an angle in a right triangle.

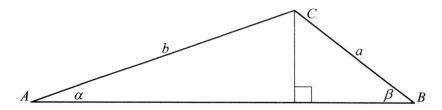

Figure 20.6   Standard proof of Law of Sines on plane

For triangles on the sphere we can find a very similar result.

> **b.** *Show that for a small triangle on a sphere with radius ρ,*
>
> $$(\sin a/\rho)/\sin \alpha = (\sin b/\rho)/\sin \beta = (\sin c/\rho)/\sin \gamma.$$

If △ACD is a triangle on the sphere with the angle at D being a right angle, then use gnomic projection to project △ACD onto the plane that is tangent to the sphere at A. Because the plane is tangent to the sphere at A, the size of the angle α is preserved under the projection. In general, angles on the sphere not at A will not be projected to angles of the same size, but in this case the right angle at D will be projected to a right angle. (*Be sure you see why this is the case. A good drawing and the use of symmetry and similar triangles will help.*) Now express the sine of α in terms of the sides of this projected triangle.

When we measure the sides in radians, on the sphere the Law of Sines becomes

$$(\sin a)/\sin \alpha = (\sin b)/\sin \beta = (\sin c)/\sin \gamma.$$

For a right triangle this becomes (Figure 20.7)

$$\sin \alpha = (\sin a)/(\sin c).$$

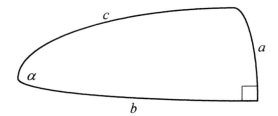

Figure 20.7   Law of Sines for right triangles on a sphere

DUALITY ON A SPHERE

We can now ask, Is it possible to find expressions corresponding to the other triangle congruence theorems that we proved in Chapters 6 and 9? Let us see how we can be helped by a certain concept of duality that we will now develop.

When we were looking at SAS and ASA, we noticed a certain duality between points and lines (geodesics). SAS was true on the plane or open hemisphere because *two points determine a unique line segment* and ASA was true on the plane or open hemisphere because *two (intersecting) lines determine a unique point*. In this section we will make this notion of duality broader and deeper and look at it in such a way that it applies to both the plane and the sphere.

On the whole sphere, two distinct points determine a unique straight line (great circle) *unless the points are antipodal*. In addition, two distinct great circles determine a unique *pair of antipodal points*. Also, a circle on the sphere has two centers *that are antipodal*. Remember also that in most of the triangle congruence theorems, we had *trouble with triangles that contained antipodal points*. So our first step is to consider not points on the sphere but rather ***point-pairs***, pairs of antipodal points. With this definition in mind, check the following:

- *Two distinct point-pairs determine a unique great circle (geodesic).*

- *Two distinct great circles determine a unique point-pair.*

- *The intrinsic center of a circle is a single point-pair.*

- *SAS, ASA, SSS, AAA are true for all triangles not containing any point-pairs.*

Now we can make the duality more definite.

- *The dual of a great circle is its* **poles** (the point-pair that is the intrinsic center of the great circle). Some books use the term *"***polar***"* in place of *"***dual***"*.

- *The dual of a point-pair is its* **equator** (the great circle whose center is the point-pair).

Note: *If the point-pair P is on the great circle* **l**, *then the dual of* l *is on the dual of P*. See Figure 20.8.

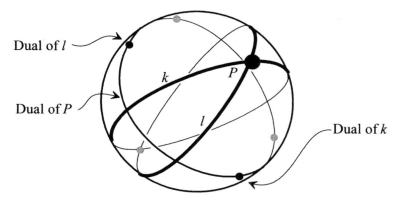

Figure 20.8   *P* is on *l* implies the dual of *l* is on the dual of *P*

If *k* is another great circle through *P*, then notice that the dual of *k* is also on the dual of *P*. Because an angle can be viewed as a collection of lines (great circles) emanating from a point, the dual of this angle is a collection of point-pairs lying on the dual of the angle's vertex. And, vice versa, the dual of the points on a segment of a great circle are great circles emanating from a point-pair that is the dual of the original great circle. *Before going on, be sure to understand this relationship between an angle and its dual.* Draw pictures. Make models.

PROBLEM 20.4   THE DUAL OF A SMALL TRIANGLE ON A SPHERE

The dual of the small triangle $\triangle ABC$ is the small triangle $\triangle A^*B^*C^*$, where A^* is that pole of the great circle of BC that is on the same side of BC as the vertex A, similarly for B^*, C^*. See Figure 20.9.

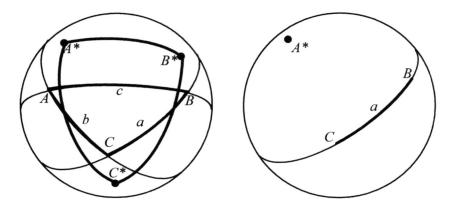

Figure 20.9   The dual of a small triangle

a. *Find the relationship between the sizes of the angles and sides of a triangle and the corresponding sides and angles of its dual.*

b. *Is there a triangle that is its own dual?*

PROBLEM 20.5 TRIGONOMETRY WITH CONGRUENCES

a. *Find a dual of the Law of Cosines on the sphere. That is, find a statement that results from replacing in the Law of Cosines every angle and every side by its dual.*

There is an analogous dual Law of Cosines for a hyperbolic plane, proved in the same books cited with Theorem 20.2, the hyperbolic Law of Cosines.

b. *For each of ASA, RLH, SSS, AAA, if you know the measures of the given sides and angles, how can you find the measures of the sides and angles that are not given? Do this for both spheres and hyperbolic planes.*

Use part **a** and the formulas from Problems **20.1** and **20.2**.

DUALITY ON THE PROJECTIVE PLANE

The gnomic projection, *g* (Problem **14.2**), allows us to transfer the above duality on the sphere to a duality on the plane. If *P* is a point on the plane, then there is a point Q on the sphere such that $g(Q) = P$. The dual of Q is a great circle l on the sphere. If Q is not the south pole, then half of l is in the southern hemisphere and its projection onto the plane, $g(l)$, is a line we can call the ***dual of P***. This defines a dual for every point on the plane except for the point where the south pole of the sphere rests. See Figure 20.10. It is convenient to call this point the origin, O, of the plane.

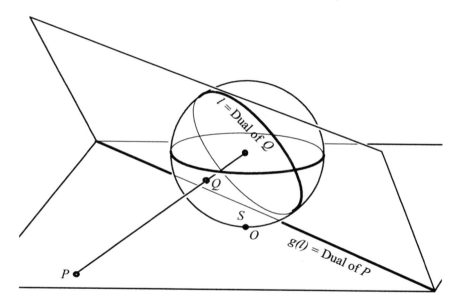

Figure 20.10   Duality on the projective plane

Note that O is the image of S, the south pole, and that the dual of S is the equator, which is projected by G to infinity on the plane. Thus we define the dual of O to be the ***line at infinity***. If l is any line in the plane, then it is the image of a great circle on the sphere that intersects the equator in a point-pair. The image of this point-pair is considered to be a *single* point at infinity at the "end" of the line l, the *same* point at both ends. The plane with the line at infinity attached is called the ***projective plane***. This projective plane is the same as the *real projective plane* that we investigated in Problem **18.3b**, the only difference being that here we are focusing on the gnomic projection onto the plane and in Chapter 18 we were considering it as a spherical 2-manifold.

*PROBLEM 20.6   PROPERTIES ON THE PROJECTIVE PLANE

a.   *Check that the following hold on the projective plane:*

♦   *Two points determine a unique line.*

♦   *Two parallel lines share the same point at infinity.*

♦   *Two lines determine a unique point.*

♦   *If a point is on a line, then the dual of the line is a point that is on the dual of the original point.*

b.   *If γ is the circle with center at the origin and with radius the same as the radius of the sphere and if (P,Q) is an inversive pair with respect to γ, then show that the dual of P is the line perpendicular to OP at the point –Q, the point in the plane that is opposite (with respect to O) of Q. See Figure 20.11.*

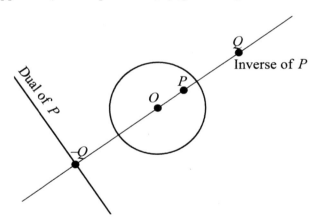

Figure 20.11 Duals/inversions

c.   *Show that the dual of a point P on γ is a line tangent to γ at the diametrically opposite point –P.*

DUAL VIEWS OF OUR EXPERIENCE

Our usual point of view for perceiving our world is from the origin (because in this view we are the center of our world). We look out in all directions to observe what is outside and we strive to look out as close toward infinity as we can. The dual of this is the point of view where we are the line at infinity (the dual of the origin), we view the whole world (which is within us), and we strain to look in as close toward the center as we can.

PERSPECTIVE DRAWING AND HISTORY

Look at the perspective drawing in Figure 20.12.

Present-day theories of projective geometry got their start from Euclid's *Optics* [**AT:** Euclid *Optics*] and later from the theories of perspective developed during the Renaissance by artists who studied the geometry inherent in perspective drawings. The first mathematical rule for getting the correct perspective was invented by Filippo Brunelleschi (1377–1455) in or shortly before 1413. In the Renaissance there was no such profession as artist. Some of the most famous artists of 14th and 15th centuries were trained as goldsmiths and were taught some mathematics from books such as Leonardo of Pisa's (c.1170–c.1240) book *Libro d'abaco*, mostly known for its use of Arabic numerals and origin of Fibonacci numbers, but there is much more mathematics in it. Brunelleschi's work included painting and sculpture as well as the design of stage machinery and buildings. He invented his method of "artificial perspective" for drawing different buildings in Rome in a way that appeared realistic. But no written descriptions survive from Brunelleschi.

The earliest works to show parallel lines that converge to a point in a more or less convincing manner are not paintings, but relief panels by a friend of Brunelleschi's, the sculptor Donatello (1386–1466). Another younger friend of Brunelleschi's who was also an early user of perspective was the painter Masaccio (1401–c.1428). Masaccio's fresco *Trinity* (c.1426) in Santa Maria Novella (Florence) provides the best opportunity for studying Masaccio's mathematics and hence making some educated guesses about Brunelleschi's method.

The first surviving written account of a method of perspective construction is by Leon Batista Alberti (1404–1472) in the beginning of a short treatise *On Painting* in 1435. The problem was how to construct the perspective image of a square checkerboard pavement. It is clear from later literary evidence, mainly dating from 16th century, that Alberti's construction was not the only one current in 15th century. Among the first mathematical treatises on perspective we should mention also Pierro della Francesca (1412–1492), *De prospectiva pingendi*. Pierro della Francesca is acknowledged as one of the most important painters of the 15th century, but he had an independent reputation as a highly competent mathematician. In Albrecht Dürer's (1471–1528) *Treatise on measurement with compasses and straightedge* (Nuremberg, 1525) is shown an instrument being used to draw a perspective image of a lute. Another mechanical aid described by Dürer goes back to the early 15th century and was described also by Alberti and Leonardo da Vinci. It was a veil of coarse netting, which was used to provide something rather like a system of coordinate lines. Such squaring systems were familiar to artists also as a method of scaling drawings up for transfer to the painting surface. For more discussion on perspective and mathematics in the Renaissance, see [**AD:** Field].

The "vanishing point" where the lines that are parallel on the box intersect on the horizon of the drawing is an image on the drawing of the point at infinity on these parallel lines.

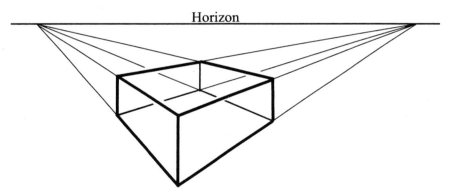

Figure 20.12   Drawing with two-point perspective

One way to visualize this is to imagine yourself at the center of a transparent sphere looking out at the world. If you look at two parallel straight lines and trace these lines on the sphere, you will be tracing segments of two great circles. If you followed this tracing indefinitely to the ends of the straight lines, you would have a tracing of two half great circles on the sphere intersecting in their endpoints. These endpoints of the semicircles are the images of the point at infinity that is the common intersection of the two parallel lines. If you now use a gnomic projection to project this onto a plane (for example, the artist's canvas), then you will obtain two straight line segments intersecting at one of their endpoints as in the drawing above.

Chapter 21

MECHANISMS

The mathematical investigations referred to bring the whole apparatus of a great science to the examination of the properties of a given mechanism, and have accumulated in this direction rich material, of enduring and increasing value. What is left unexamined is however the other, immensely deeper part of the problem, the question: How did the mechanism, or the elements of which it is composed, originate? What laws govern its building up?

— F. Reuleaux, *Kinematics* (1876), p. 3

In this chapter we will study *mechanisms*, which for our purposes we define as collections of rigid bodies with moveable connections having the purpose of transforming motion. We have already studied two mechanisms: an angle-trisecting mechanism in Problem **15.4d** and the Peaucellier-Lipkin straight-line mechanism in Problem **16.3**. A *machine* can be considered as a combination of mechanisms connected together in a way to do useful work.

In this chapter we will use the Law of Cosines and the Law of Sines for the plane and sphere in Problems **21.1** and **21.2**. Otherwise the material demands only basic understanding of plane and spheres.

INTERACTIONS OF MECHANISMS WITH MATHEMATICS

We mentioned in Chapter 0 that one of the strands in the history of geometry is the Motion/Machines Strand; and we showed in Chapter 1 how this strand led to mechanism for producing straight line motion. It is known that the Greeks, in particular Aristotle, studied the so-called simple machines: the wheel, lever, pulleys, and inclined plane. He also described gear wheel drive in windlasses and pointed out that the direction of rotation is reversed when one gear wheel drives another gear wheel. Archimedes made devices to multiply force or torque many times and studied spirals and helices for mechanical purposes (see Figure 0.5). We know Hero's formula for the area of a triangle but in his time he was better known as an engineer. Most of his inventions are known from the writings of the Roman engineer Vitruvius. However, there was little application of Greek natural science to engineering in antiquity. In fact, engineering contributed far more to science than science did to engineering until the latter half of the 19th century. For more discussion of this history, see [**ME**: Kirby], p. 43.

One of the simplest mechanisms used in human activities are linkages. Perhaps an idea of using linkages came into somebody's mind because a linkage resembles a human arm. We can find linkages in old drawings of various machines in 13th century; see Figure 1.3. Leonardo da Vinci's *Codex Madridi* (1493) contained a collection of machine elements, *Elementi Macchinali*, and he invented a lathe for turning parts with elliptical cross section, using a four-bar linkage (see Problem **21.1**). Georgius Agricola (1494–1555) is considered a founder of geology as a discipline but he gave descriptions of machines used in mining, and that is where we can find pictures of linkages used in these machines. Gigantic linkages, principally for mine pumping operations, connected water wheels at the riverbank to pumps high up on the hillside. Such linkages consisted mostly of what we call four-bar linkages; see Problem **21.1**.

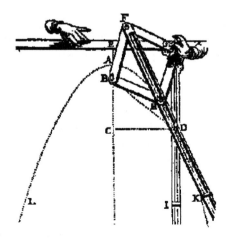

Figure 21.1 Mechanism for drawing a parabola (1657)

Mathematicians got interested in linkages first for geometric drawing purposes. We know about some such devices from ancient Greek mathematics. (For example, see Problem **15.4** about devices for trisecting an angle.) When Rene Descartes published his *Geometry* (1637) he did not create a curve by plotting points from an equation. There were always first given geometrical methods for drawing each curve with some apparatus, and often these apparatus were linkages. See, for example, Figure 21.1, which depicts the mechanism for drawing a parabola that appear in the works of Franz von Schooten (1615–1660) that were a popularization of Descartes' work.

Isaac Newton developed mechanisms for the generation of algebraic curves of the third degree. This tradition of seeing curves as the result of geometric actions can be found also in works of Roberval, Pascal, and Leibniz. Mechanical devices for drawing curves played a fundamental role in creating new symbolic languages (for example, calculus) and establishing their viability. The tangents, areas, and arc length associated with many curves were known before any algebraic equations were written. Critical experiments using curves allowed for the coordination of algebraic representations with independently established results from geometry. For more detailed discussion of the ideas in the last two paragraphs, see [**HI**: Dennis], Chapter 2.

Linkages are closely related with kinematics or geometry of motion. First it was the random growth of machines and mechanisms under the pressure of necessity. Much later, algebraic speculations on the generation of curves were applied to physical problems.

Two great figures appeared in 18th century, Leonard Euler (1707–1783) and James Watt (1736–1819). Although their lives overlap there was no known contact between them. But both of them were involved with the "geometry of motion." Watt, instrument maker and engineer, was concerned with designing mechanisms that produce desired motions. Watt's search for a mechanism to relate circular motion with straight-line motion is discussed in the historical introduction to Chapter 1. Euler's theoretical results were unnoticed for a century by the engineers and mathematicians who were devising linkages to compete or supersede Watt's mechanism. See Figure 1.4.

The fundamental idea of the geometric analysis of motion (kinematics) stems from Euler, who wrote in 1775,

> The investigation of the motion of a rigid body may be
> conveniently separated into two parts, the one geometrical, the

other mechanical. In the first part, the transference of the body from a given position to any other position must be investigated without respect to the causes of motion, and must be represented by analytical formulae, which will define the position of each point of the body. This investigation will therefore be referable solely to geometry ... . [Euler, *Novi commentarii Academiae Petrop.*, vol. XX, 1775. Translation in Willis, *Principles of Mechanism*, 2nd ed., p. viii, 1870.]

These two parts are sometimes called kinematics (geometry of motion) and kinetics (the mechanics of motion). Here we can see beginnings of the separation of the general problem of dynamics into kinematics and kinetics.

Franz Reuleaux (1829–1905) divided the study of machines into several categories and one of them was study of the geometry of motion ([**ME**: Reuleaux], pp. 36–40). In the proliferation of machines at the height of the Industrial Revolution, Reuleaux was systematically analyzing and classifying new mechanisms based on the way they constrained motion. He hoped to achieve a logical order in engineering. The result would be a library of mechanisms that could be combined to create new machines. He laid the foundation for a systematic study of machines by determining the basic building blocks and developing a system for classifying known mechanism types. Reuleaux created at Berlin a collection of over 800 models of mechanisms and authorized a German company, Gustav Voigt, *Mechanische Werkstatt*, in Berlin, to manufacture these models so that technical schools could use them for teaching engineers about machines.

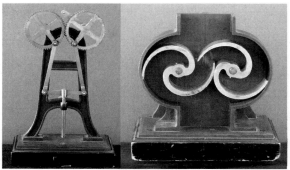

Figure 21.2 Three Reuleaux models

In 1882, Cornell University acquired 266 of such models, and now the remaining 219 models is the largest collection of Reuleaux kinematic mechanisms in the world. See examples in Figure 21.2. With support from the NSF, a team of Cornell mathematicians, engineers, and librarians has developed a digital Reuleaux kinematic model Web site [**ME**: KMODDL] as the part of National Science Digital Library. The publicly available Web site contains photos, mathematical descriptions, historical descriptions, moving virtual reality images, simulations, learning modules (for middle school through undergraduate), and even downloadable files for 3D printing. In this chapter we will look at the mathematics related to a few of these mechanisms in order to show geometry in Machine/Motion Strand.

There will be discussions of more recent history later in the chapter.

PROBLEM 21.1 FOUR-BAR LINKAGES

A four bar linkage is a mechanism that lies in a plane (or spherical surface) and consists of four bars connected by joints that allow rotation only in the plane (or sphere) of the mechanism. See Figure 21.3.

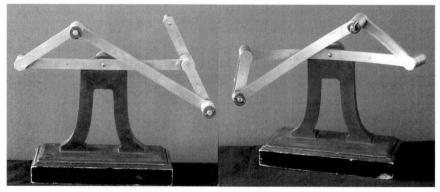

Planar   Spherical
Figure 21.3   Reuleaux four-bar linkages

In normal practice one of the links is fixed so that it does not move. In Figure 21.4 we assume that the link *OC* is fixed and investigate the possibilities of motion for the other three links. We call the link *OA* the *input crank* and link *CB* the *output crank*. Similarly, we call the angle θ the *input angle* and angle ϕ the *output angle*.

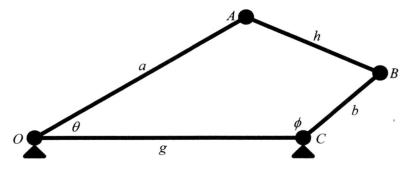

Figure 21.4 Four-bar linkage

a. (Plane) *The input crank will be able to swing opposite C (where the input angle is $\pi = 180$ degrees) only if $a + g < b + h$. If $a + g > b + h$, there will be a maximum input angle, θ_{max}, satisfying*

$$\cos \theta_{max} = \frac{(g^2 + a^2) - (h + b)^2}{2ag}.$$

What happens if $a + g = b + h$?

You may find it helpful to experiment with four-bar linkages that you make out of strips of cardboard; and/or to play with the online simulations that are linked on the Experiencing Geometry Web page. Apply the Law of Cosines (Problem **20.2**).

We can now do the same analysis on the sphere, using the spherical Law of Cosines (**20.2**) and the special absolute value $|l|_s$ (see the section Triangle Inequality just before Problem **6.3**). On the plane, $|l|_s = |l| =$ length of l. On the sphere, $|l|_s =$ (shortest) distance between the endpoints of l.

b. (Sphere) *The input crank will be able to swing opposite C (where the input angle is $\pi = 180$ degrees) only if $|a + g|_s < b + h$. If $|a + g|_s > b + h$, there will be a maximum input angle, θ_{max}, satisfying*

$$\cos \theta_{max} = \frac{\cos(h + b) - \cos a \cos g}{\sin a \sin g},$$

where the edge lengths are measured by the radian measure of the angle they subtend at the center of the sphere.

What happens if $|a + g|_s = b + h$?

We can now finish this problem working with both the plane and the sphere at the same time.

    c. (Plane or Sphere) *Similarly, whenever $|b-h| > |g-a|$ there will be a minimum input angle, θ_{min}, satisfying*

$$\cos \theta_{min} = \frac{g^2 + a^2 - |b-h|^2}{2ag} \text{ (plane)},$$

$$\cos \theta_{min} = \frac{\cos|b-h| - \cos a \cos g}{\sin a \sin g} \text{ (sphere)}.$$

The minimum angle is actualized when either B-A-C is straight or A-C-B is straight, as in Figure 21.5.

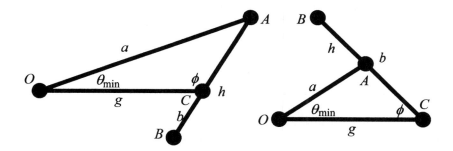

Figure 21.5 Minimum input angles

Thus we have four types of input cranks:

1. A ***crank*** if the link *OA* can freely rotate completely around *O*. In this case, $b + h > |a + g|_s$ and $|a - g| > |b - h|$.

2. A ***0-rocker*** if there is a maximum input angle but the link *OA* can rotate freely past $\theta = 0$. Then $b + h < |a + g|_s$ and $|a - g| > |b - h|$.

3. A ***π-rocker*** if there is a minimum input angle but the link *OA* can rotate freely past $\theta = \pi$. Then $b + h > |a + g|_s$ and $|a - g| < |b - h|$.

4. A ***rocker*** if there is a maximum input angle and minimum input angle. In this case, $b + h < |a + g|_s$ and $|a - g| < |b - h|$.

The analysis of the output crank is exactly symmetric to above with the lengths *a* and *b* interchanged. In particular,

    d. *If $|b + g|_s > a + h$, then there is a maximum output angle ϕ_{max} that satisfies*

$$\cos\theta_{max} = \frac{(g^2 + b^2) - (h + a)^2}{2ag} \quad \text{(plane)}$$

$$\cos\theta_{max} = \frac{\cos(h + a) - \cos b \cos g}{\sin b \sin g} \quad \text{(sphere)}.$$

If $|a - h| > |g - b|$, there is a minimum output angle ϕ_{min} that satisfies

$$\cos\theta_{min} = \frac{g^2 + b^2 - |a - h|^2}{2bg} \quad \text{(plane)},$$

$$\cos\theta_{min} = \frac{\cos|a - h| - \cos b \cos g}{\sin b \sin g} \quad \text{(sphere)}.$$

Thus we have four types of output cranks:

1. A **crank** if the link CB can freely rotate completely around C. In this case, $a + h > |b + g|_s$ and $|b - g| > |a - h|$.

2. A **0-rocker** if there is a maximum output angle but link CB can rotate freely past $\phi = 0$. Then $a + h < |b + g|_s$ and $|b - g| > |a - h|$.

3. A **π-rocker** if there is a minimum output angle but the link CB can rotate freely past $\phi = \pi$. Then $a + h > |b + g|_s$ and $|b - g| < |a - h|$.

4. A **rocker** if there is a maximum output angle and minimum output angle. In this case, $a + h < |b + g|_s$ and $|b - g| < |a - h|$.

    e. *Putting these together we get eight types of four-bar linkages. Make a model of each type using cardboard strips.*

1. A **double crank** in which the input and output links are cranks. $b + h > |a + g|_s$, $|a - g| > |b - h|$, $a + h > |b + g|_s$, and $|b - g| > |a - h|$.

2. A **crank-rocker** if the input link is a crank and the output link is a rocker. $b + h > |a + g|_s$, $|a - g| > |b - h|$, $a + h < |b + g|_s$, and $|b - g| < |a - h|$.

3. A **rocker-crank** if the input link is a rocker and the output link is a crank. $b + h < |a + g|_s$, $|a - g| < |b - h|$, $a + h > |b + g|_s$, and $|b - g| > |a - h|$.

4. A **rocker-rocker** if both the input and the output link are rockers. $b + h < |a + g|_s$, $|a - g| < |b - h|$, $a + h < |b + g|_s$, and $|b - g| < |a - h|$.

5. A **00 double rocker** if there are maximum input and output angles and no minimums, thus both cranks move across the fixed OC. $b + h < |a + g|_s$, $|a - g| > |b - h|$, $a + h < |b + g|_s$, and $|b - g| > |a - h|$.

6. A *0π double rocker* if the input angle has a maximum and no minimum but the output angle has a minimum but no maximum. $b + h < |a + g|_s$, $|a - g| > |b - h|$, $a + h > |b + g|_s$, and $|b - g| < |a - h|$.

7. A *π0 double rocker* if the input angle has a minimum and no maximum but the output angle has a maximum but no minimum. $b + h > |a + g|_s$, $|a - g| < |b - h|$, $a + h < |b + g|_s$, and $|b - g| > |a - h|$.

8. A *ππ double rocker* if the input and output angles both have minimums but no maximums and move freely on the ends of *OC*. $b + h > |a + g|_s$, $|a - g| < |b - h|$, $a + h > |b + g|_s$, and $|b - g| < |a - h|$.

Check that the other eight combinations (a 0 or π rocker combined with a crank or rocker) are not possible. This can be done either analytically (using the inequalities) or geometrically (by noting symmetries).

All the other four-bar linkages are the cases when one or more of the inequalities become equalities, in each of these cases the linkage can be folded. That is, the linkage has a configuration in which all the links line up with *OC*. Some four-bar linkages can be folded in more than one way; for example, the linkage with $a = h = b = g$ can be folded in three different ways (*Try it!*).

PROBLEM **21.2** UNIVERSAL JOINT

Almost all vehicles with an engine in front that drives the rear wheels have a drive shaft that transmits the power from the engine to the rear axle. It is important that the drive shaft be able to bend as the vehicle goes over bumps. The usual way to accomplish this "bending" is to put in the drive shaft a *universal joint* (also known as Hooke's joint or Cardan's joint). See Figure 21.6. To see this model in motion, go to the Experiencing Geometry Web page, which will link to VR-movies on [**ME:** KMODDL].

In 1676, Robert Hooke (1635–1703) published a paper on an optical instrument that could be used to study the sun safely. In order to track the sun across the sky, the device featured a control handle fitted with a new type of joint that allowed twisting motion in one shaft to be passed on to another, no matter how the two shafts were orientated. Hooke gave this the name "universal joint." This joint was earlier suggested by Leonardo da Vinci and also is attributed to Girolamo Cardano. Therefore, on the Euorpean continent it got name "Cardan's joint," but in Britain the name of "Hooke's joint" was used.

Figure 21.6   Universal joint

a. *The universal joint can be considered to be a spherical four-bar linkage with a = b = h = π/2. The fixed (grounded) link g is the angle between the input and output shafts that can be adjusted in the range π/2 < g ≤ π.*

The links *a*, *h*, and *g* are not actually links; however, the constraints of the mechanism operate as if they were spherical links. See Figure 21.7.

b. *For what lengths of the fixed link (angles between the input and output shafts) is the universal joint a double crank (see **21.1e**)?*

This is of utmost importance in its automotive use because as an automobile goes over bumps the angle *g* between the shafts will change.

We now look at the relationship between the rotation of the input shaft to the rotation of out shaft. One of the problems with the universal joint is that, though one rotation of the input shaft results in one rotation of the output shaft, the rotations are not in sync during the revolution.

c. *Check that Figure 21.7 is correct. In particular,*

i. *A is the pole for the great circle (dashed in the figure) passing through O and B; and the great circle arcs h and a intersect this great circle at right angles.*

ii. *Likewise, B is the pole for the great circle passing through A and C; and the great circle arcs h and b also intersect this great circle at right angles.*

iii. *The arc h must bisect the lune determined by these two dashed great circles. The angle of this lune at P must be π/2 as marked.*

iv. *If we use α to label the angle at A and β to label the angle at B, as in the figure, then radian measure of the arc OB is α, and the radian measure of the arc AC is β.*

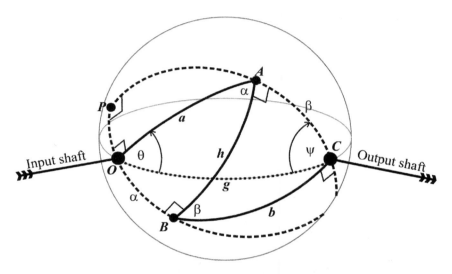

Figure 21.7 Universal joint as a spherical mechanism

The angle θ measures the rotation of the input shaft and the angle ψ measures the rotation of the output shaft. Note that, when $\theta = 0$ then $\psi = 0$, also. As θ changes in the positive counterclockwise direction, ψ will be changing in the negative clockwise direction; thus, when θ is positive, ψ will be negative and we will need to use $-\psi$ when denoting the angle in the triangle *OCA*.

d. *The input and output angles satisfy*

$$\tan(-\psi) = \frac{\tan\theta}{-\cos g}.$$

In practice, g is near π and $-\cos g$ is positive. Note that when g = π, $\tan(-\psi) = \tan\theta$ and thus the two shafts turn in unison.

Hint: Apply the Spherical Pythagorean Theorem (Theorem **20.2a**) to the right triangle *OPC*, and the Law of Sines (Problem **20.3b**) to triangles *OCA* and *OCB*.

Problem 21.3 Reuleaux Triangle and Constant Width Curves

What is this triangle? If an enormously heavy object has to be moved from one spot to another, it may not be practical to move it on wheels. Instead the object is placed on a flat platform that in turn rests on cylindrical rollers. As the platform is pushed forward, the rollers left behind are picked up and put down in front. An object moved this way over flat horizontal surface does not bob up and down as it rolls along. The reason is that cylindrical rollers have a circular cross section, and a circle is closed curve *with constant width*. What does this mean? If a closed convex curve is placed between two parallel lines and the lines are moved together until they touch the curve, the distance between the parallel lines is the curve's width in one direction. Because a circle has the same width in all directions, it can be rotated between two parallel lines without altering the distance between the lines.

Reuleaux model   Wankel engine
Figure 21.8 Mechanisms utilizing a Reuleaux triangle

Is the circle the only curve with constant width? Actually there are infinitely many such curves. The simplest noncircular such curve is named the Reuleaux triangle. Mathematicians knew it earlier (some authors refer to Leonard Euler in 18th century), but Reuleaux was the first to demonstrate and use of its constant width properties. In Figure 21.8 there is a Reuleaux model using the Reuleaux triangle and an image of inside of a Wankel engine (similar to that used in some Mazda automobiles) showing the rotor in the shape of a Reuleaux triangle.

A Reuleaux triangle can be constructed starting with an equilateral triangle of side s and then replacing each side by a circular arc with the other two original sides as radii. See Figure 21.9.

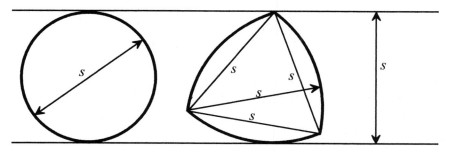

Figure 21.9 Circle and Reuleaux triangle of same width

a. *Why will the Reuleaux triangle make a convenient roller but not a convenient wheel?*

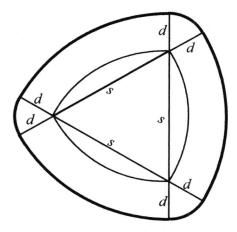

Figure 21.10 Smoothed Reuleaux triangle

The Reuleaux triangle has corners, but if you want to smooth out the corners you can extend a Reuleaux triangle a uniform distance d on every side as in Figure 21.10. Then you can

b. *Show that the resulting curve has constant width $s + 2d$.*

Figure 21.11 Constant width coins

Other symmetrical curves of constant width result if you start with a regular pentagon (or any regular polygon with an odd number of sides) and follow similar procedures. See Figure 21.11 for examples of British coins that are the shape of a constant width curve based on the heptagon. *What advantages would these British coins have due to their shape?*

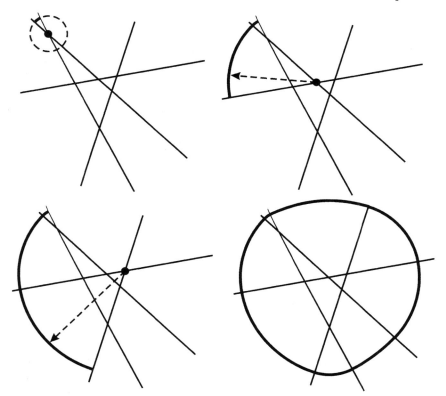

Figure 21.12 Creating irregular curves of constant width

But here is one really surprising method of constructing curves with constant width: Draw as many straight lines as you please such that each line intersects all the others. See Figure 21.12. On one of the lines start with a point sufficiently far away from the intersections. Now draw an arc from this point to an adjacent line, with the compass point at the intersection of the two lines. Then, starting from the end of this arc, draw another arc connecting to the next line with the compass point at the intersection of these two lines. Proceed in this manner from one line to the next, as indicated in Figure 21.12.

If you do it carefully, the curve will close and will have a constant width. (You can try to prove it! It is not difficult at all.) The curves drawn in this way may have arcs of as many different circles as you wish. The example in Figure 21.12 shows steps in drawing such curves, but you will really enjoy making your own. After you have done that, you can make several more copies of it and check that your wheels really roll!

c. *Prove that the procedure in the last paragraph produces a curve of constant width as long as all the intersection points are inside the curve. Can you specify how far out you have to start in order for this to happen?*

The Reuleaux triangle has been used to make a drill bit that will drill a (almost) square hole. See Figure 21.13.

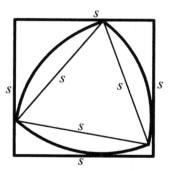

Figure 21.13 Reuleaux triangle in a square

d. *A Reuleaux triangle of width s can be turned completely around within a square of side s in such a way that, at each time of the motion, the Reuleaux triangle will be tangent simultaneously to* *all four sides of the square. Describe the small spaces in the corners of the square that the Reuleaux triangle will not reach.*

The interested reader may find more information and references about Reuleaux triangles in [**ME**: KMODDL]; go to the Experiencing Geometry Web page for direct links.

To sharpen your interest, we list some further properties of Reuleaux triangles:

1. The inscribed and circumscribed circles of an arbitrary curve of constant width h are concentric and the sum of their radii is equal to h.

2. Among curves with constant width h, the circle bounds the region of greatest area and the Reuleaux triangle bounds the region of least area.

3. Any curve of constant width h has perimeter equal πh.

4. The corners of a Reuleaux triangle are the sharpest possible on a curve with constant width.

5. The circle is the only curve of constant width with central symmetry (see page 15).

6. For every point on a curve of constant width there exists another point with the distance between the two equaling the width of the curve, and the line joining these two points is perpendicular to the support lines at both points. (***Support lines*** are lines that touch the curve and the curve lies totally on one side of the line.)

7. There is at least one supporting line through every point of a curve of constant width.

8. If a circle has three (or more) points in common with a curve of constant breadth h, then the length of the radius of the circle is at most h.

INVOLUTES

Reuleaux used the geometric idea of involutes in the design of several mechanisms including a pump (the middle image in Figure 21.2) and gear design (Figure 21.15). Before we discuss these we must describe the geometry of the involute. Focus on one of the four arms in the pump in Figure 21.14.

Figure 21.14 Involute arm

Imagine that the outlined circle in picture is a spool that has a black thread with white edging wrapped around it in such a way that when fully wound the end of the thread is at the point *A*. Now imagine unwinding the thread, keeping the spool fixed, and keeping the thread pulled taut. The end of the thread traces the outer edge of the spiral arm. This curve is called the *involute of a circle* (in this case, the outlined circle).

Instead of keeping the spool fixed and unwinding the thread, we could rotate the spool and pull the thread taut in the same direction. It is this latter view that we will use in analyzing the spiral pump.

Examine the picture of the spiral pump in Figure 21.15. Now imagine that there is thread rolled around the left spool and then pulled taut and wrapped around the right spool as indicated in the picture. Place a small circle on the thread at the place that two arms touch each other. Now, instead of unwrapping the thread, we will turn the two spools at the same rate, always keeping the thread taut. Since the spiral arms are in the shape of an involute, the small circle will follow the outer edge of both spiral arms. Thus, as the spools rotate, the spiral arms will stay in contact.

Figure 21.15 Spiral pump with thread and small circle

The two rotors in the spiral pump can be thought of as gears with two teeth. The same discussion above illustrates why it is advantageous for gear teeth, in general, to be in the shape of an involute curve so that the gear teeth stay in contact throughout uniform turning of the gears. If the axes of two engaging gears are parallel, then the involute is a planar involute, as described above. However, if the axes are not parallel, then the gears can be considered as on a sphere whose center is at the intersection of the two axes. An involute of a circle (on plane or sphere) can be described either as unrolling a taut string from the circle or as rolling of a straight line along the circle.

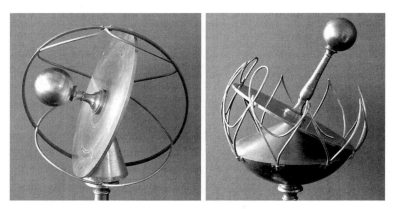

Figure 21.16 Spherical involute curves

Reuleaux designed several models to illustrate spherical involutes by rolling a straight line (great circle) on a small circle. See Figure 21.16 for photos of these models, but more understanding will be possible if you experience the VR-movies on [**ME**: KMODDL], see direct links on the Experiencing Geometry Web page.

LINKAGES INTERACT WITH MATHEMATICS

After Descartes and others used linkages to draw curves, it was natural for mathematicians to ask the question of what curves could be drawn by linkages. In [**ME**: Kempe 1876], A. B. Kempe gave a proof that any algebraic curve may be described by a linkage. The idea for Kempe's proof, as discussed in [**ME**: Artobolevski], is as follows:

Consider the algebraic curve $f(x, y) = 0$, which can be expressed in the form $\sum A_{mn} x^m y^n = 0$, where the coefficients A_{mn} are constant. Thus the generation of the curve reduces to a series of mathematical operations. Kempe's idea was that each of these mathematical operations can be fulfilled by individual linkages, which can then be linked together into a kinematic chain of linkages. Linkages needed for this are as follows:

a. Linkage for translating a point along a given straight line (for example, the Peaucellier-Lipkin linkage in Problem **16.3**);

b. Linkage for projecting a given point onto a given line;

c. Linkage that cuts off on one axis a segment equal to a given segment on the other axis;

d. Linkage that determines a straight line that passes through a given point and is parallel to a given line;

e. Linkage that, given two segments r and s on one line and one segment t on another line, will obtain a second segment u on the second line such that $r/s = t/u$ (Multiplier);

f. Linkage for the addition of two given segments (Adder).

For more details on this proof, see [**ME**: Artobolevski]. For a different proof with more explicit pictures of the constructions of linkages and their combinations, see [**ME**: Yates], Section 11.

In a different direction, there was for many years an open question that appeared in robotics, topology, discrete geometry, and pattern recognition:

*Given a linear **chain** of links* (each one connected to the next to form a polygonal path without self-intersections) *or a **cycle** of links* (a linear chain with the first and last links joined*), then is it possible to find a motion of the chain or cycle during which there continue to be no self-intersections and, at the end of the motion, the chain forms a straight line and the cycle forms a convex polygon.* See Figure 21.17.

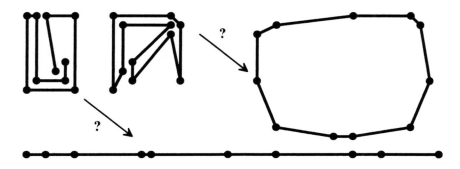

Figure 21.17   Straightening and convexifying linkages

In 2003, Robert Connelly, Erik D. Demaine, and Günter Rote published a paper, "Straightening Polygonal Arcs and Convexifying Polygonal Cycles" [**ME**: Connelly], in which they solved this problem positively and, in addition, proved that their motion is piecewise differentiable, does not decrease the distance between any pair of vertices, and preserves any symmetry present in the initial configuration.

Chapter 22

3-SPHERES AND HYPERBOLIC 3-SPACES

> Let us, then, make a mental picture of our universe: ... as far as possible, a complete unity so that whatever comes into view, say the outer orb of the heavens, shall bring immediately with it the vision, on the one plane, of the sun and of all the stars with earth and sea and all living things as if exhibited upon a transparent globe.
>
> Bring this vision actually before your sight, so that there shall be in your mind the gleaming representation of a sphere, a picture holding all the things of the universe ... . Keep this sphere before you, and from it imagine another, a sphere stripped of magnitude and of spatial differences; cast out your inborn sense of Matter, taking care not merely to attenuate it: call on God, maker of the sphere whose image you now hold, and pray Him to enter. And may He come bringing His own Universe ... .
>
> — Plotinus, *The Enneads*, **V.8.9**, Burdette, NY: Larson, 1992

In this chapter you will explore hyperbolic 3-space and the 3-dimensional sphere that extrinsically sits in 4-space. But intrinsically, if we zoom in on a point in a 3-sphere or a hyperbolic 3-space, then locally the experience of the space will become indistinguishable from an intrinsic and local experience of Euclidean 3-space. This is also our human experience in our physical universe. We will study these 3-dimensional spaces both because they are possible geometries for our physical universe and in order to see that these geometries are closely related to their 2-dimensional versions.

The starred problems will require some experience with abstract vector spaces but can be skipped as long as you experience Problem **22.1** and at least assume the results of Problem **22.6**.

Try to imagine the possibility of our physical universe being a 3-sphere in 4-space. It is the same kind of imagination a 2-dimensional (2-D) being would need in order to imagine that it was on 2-sphere (ordinary sphere) in 3-space. In Problem **18.6** we thought about how a 2-D bug could determine (intrinsically) that it was on a 2-sphere. Now, we want to explore how the bug could *imagine* the 2-sphere in 3-space, that is, how could the bug imagine an extrinsic view of the 2-sphere in 3-space. In Problems 22.2 and 22.3 we will use linear algebra to help us talk about and analyze the 3-sphere in 4-space, but this will not solve the problem of imagining the 3-sphere in 4-space.

PROBLEM 22.1   EXPLAIN 2-SPHERE IN 3-SPACE TO A 2-DIMENSIONAL BUG

How would you explain a 2-sphere in 3-space to a 2-D bug living in a (Euclidean) plane or on a 2-sphere so large that it appears flat to the 2-D bug? Figure 22.1.

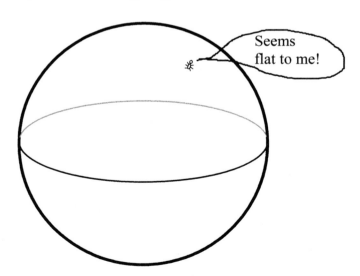

Figure 22.1   2-D bug on a large 2-sphere

SUGGESTIONS

This bug's 2-dimensional experience is very much like the experience of an insect called a water strider that we talked about in Chapter 2. A water strider walks on the surface of a pond and has a very 2-dimensional perception of the universe around it. To the water strider, there is no up or down; its whole universe consists of the surface of the water. Similarly, for the 2-D bug there is no front or back; the entire universe is the 2-dimensional plane.

Living in a 2-D world, the bug can easily understand any notions in 2-space, including plane, angle, distance, perpendicular, circle, and so forth. You can assume the bug is smart and has been in geometry class.

A bug living in a 2-D world cannot directly experience three dimensions, just as we are unable to directly experience four dimensions. Yet, with some help from you, the 2-D bug can begin to imagine three dimensions just as we can imagine four dimensions. One goal of this problem is to try to gain a better understanding of what our experience in our imagination of 4-space might be. Think about what four dimensions might be like, and you may have ideas about the kinds of questions the 2-D bug will have about three dimensions. You may know some answers, as well. The problem is finding a way to talk about them. Be creative!

One important thing to keep in mind is that it is possible to have *images* of things we cannot see. For example, when we look at a sphere, we can see only roughly half of it, but we can and do have an image of the entire sphere in our minds. We even have an image of the inside of the sphere, but it is impossible to actually see the entire inside or outside of the sphere all at once. Another example — sit in your room, close your eyes, and try to imagine the entire room. It is likely that you will have an image of the entire room, even though you can never see it all at once. Without such images of the whole room it would be difficult to maneuver around the room. The same goes for your image of the whole of the chair you are sitting on or this book you are reading.

Assume that the 2-D bug also has images of things that cannot be seen in their entirety. For example, the 2-D bug may have an image of a circle. Within a 2-dimensional world, the entire circle cannot be seen all at once; the 2-D bug can only see approximately half of the outside of the circle at a time and cannot see the inside at all unless the circle is broken. See Figure 22.2.

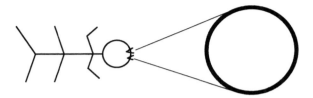

Figure 22.2   The 2-D bug sees a circle

However, from our position in 3-space we *can* see the entire circle including its inside. Carrying the distinction between what we can see and what we can imagine one step further, the 2-D bug cannot see the entire circle but can imagine in the mind the whole circle including inside and out. Thus, the 2-D bug can only imagine what we, from three dimensions, can directly see. So the 2-D bug's image of the entire circle is as if it were being viewed from the third dimension. It makes sense, then, that the image of the entire sphere that we have in our minds is a 4-D view of it, as if we were viewing it from the fourth dimension.

When we talk about the fourth dimension here, we are not talking about time, which is often considered the fourth dimension. Here, we are talking about a fourth *spatial* dimension. A fuller description of our universe would require the addition of a time dimension onto whatever spatial dimensions one is considering.

Try to come up with ways to help the 2-D bug imagine a 2-sphere in 3-space. It may help to think of intersecting planes rotating with respect to each other: How will a 2-D bug in one of the planes experience it? Draw on the bug's experience living in two dimensions as well as some of your own experiences and attempts to imagine four dimensions.

*Wʜᴀᴛ Iꜱ 4-Sᴘᴀᴄᴇ? Vᴇᴄᴛᴏʀ Sᴘᴀᴄᴇꜱ ᴀɴᴅ Bᴀꜱᴇꜱ

We could think of 4-space as $\mathbf{R}^4$:

Let $\mathbf{R}^4$ be the collection of 4-tuples of real numbers (x, y, z, w) with the *distance function* (*metric*)

$$d(\,(a,b,c,d),(e,f,g,h)\,) \equiv \sqrt{(a-e)^2 + (b-f)^2 + (c-g)^2 + (d-h)^2}$$

and *dot product* $(a,b,c,d) \cdot (e,f,g,h) \equiv ae + bf + cg + de.$

But this would be awkward sometimes because it fixes a given coordinate system; and we will find it geometrically useful to be able to change coordinates (or basis) to fit a particular problem. We will find it more powerful to have a description of space without given coordinates.

So instead of $\mathbf{R}^4$ we will think of 4-space as an (abstract) *vector space*. A vector space has a point we call the origin **O**. Then designate every point *P* by a directed straight line segment **v** from **O** to *P*. These directed line segments are, of course, what we call *vectors*. We consider **O** itself also as a vector. If you have studied vector spaces only algebraically, then it may be difficult to see the geometric content. We assume that we can *add vectors* by geometrically defining **u** + **v** to be the diagonal of the parallelogram determined by **v** and **u**, as in Figure 22.3. In order for this definition to make sense we need to assume that we can form the parallelograms. It is also possible to not identify an origin and create what is called an *affine space*. You can read about geometric affine spaces in [**DG**: Henderson], Appendix A, or [**DG**: Dodson & Poston].

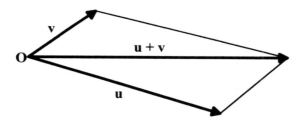

Figure 22.3 Adding vectors

We assume that addition of vectors satisfies, for all **u** and **v** in *V*,

1. **u** + **v** is in *V*, and **O** + **u** = **u**;

2. **u** + **v** = **v** + **u**, and **u** + (**v** + **w**) = (**u** + **v**) + **w**; and

3. there is a −**v** in *V* such that **v** + (−**v**) = **O**.

Geometrically, we choose a segment whose length we designate as the unit length 1. With this unit length we can determine the length of any vector **v** — we denote its length by |**v**|. For *r* a real number and **v** a vector, we define geometrically the *multiplication by scalars* *r***v** as the vector with length *r*|**v**| and lying on the straight line determined by **v**.

This multiplication by scalars satisfies, for all **u** and **v** in V and all reals r and s,

4. $r\mathbf{u}$ is in V, and $0\mathbf{u} = \mathbf{O}$, and $1\mathbf{u} = \mathbf{u}$;

5. $(r + s)\mathbf{u} = r\mathbf{u} + s\mathbf{u}$, and $r(\mathbf{u} + \mathbf{v}) = r\mathbf{u} + r\mathbf{v}$; and

6. $(rs)\mathbf{u} = r(s\mathbf{u})$.

We also assume that we can find the angle θ, $0 \le \theta \le \pi$, between any two vectors and we define the **_Euclidean inner product_** (sometimes called **_the standard inner product_**) of two vectors to be:

$\langle \mathbf{v}, \mathbf{w} \rangle = |\mathbf{v}|\,|\mathbf{w}| \cos \theta$, where θ is the angle between **v** and **w**.

Note that $\langle \mathbf{v}, \mathbf{w} \rangle$ is negative when $\theta > \pi/2$. We can check that this inner product satisfies the following properties, for all **u**, **v**, **w** in V and all reals r,

7. $\langle \mathbf{v}, \mathbf{w} \rangle = \langle \mathbf{w}, \mathbf{v} \rangle$;

8. $r\langle \mathbf{v}, \mathbf{w} \rangle = \langle r\mathbf{v}, \mathbf{w} \rangle = \langle \mathbf{v}, r\mathbf{w} \rangle$, for all reals r;

9. $\langle \mathbf{v+u}, \mathbf{w} \rangle = \langle \mathbf{v}, \mathbf{w} \rangle + \langle \mathbf{u}, \mathbf{w} \rangle$; and

10. $\langle \mathbf{v}, \mathbf{v} \rangle \ge 0$.

In a more abstract setting we could simply *define* a **_vector space with inner product_** to be a set with {an origin (**O**), addition of vectors (+), multiplication by scalars, and real-valued function $(\langle\ ,\ \rangle)$} that satisfies the 10 properties above. Then we define

$$|\mathbf{v}| = \sqrt{\langle \mathbf{v}, \mathbf{v} \rangle} \quad \text{and} \quad \cos \theta = \frac{\langle \mathbf{v}, \mathbf{w} \rangle}{|\mathbf{v}|\,|\mathbf{w}|}, \; 0 \le \theta \le \pi.$$

Note that $\langle \mathbf{v}, \mathbf{w} \rangle = 0$ implies that **v** and **w** are perpendicular.

A **_subspace_** of a vector space V is a subset $U \subset V$ if $r\mathbf{u} + s\mathbf{v}$ is in U, for every pair of vectors **u**, **v** in U and all real numbers r, s. If $\{\mathbf{u}_1, \mathbf{u}_2, ..., \mathbf{u}_n\}$ is a finite collection of vectors from V, then we call the **_span of_** $\{\mathbf{u}_1, \mathbf{u}_2, ..., \mathbf{u}_n\}$, denoted by $\text{sp}\{\mathbf{u}_1, \mathbf{u}_2, ..., \mathbf{u}_n\}$, the smallest subspace of V containing each of $\mathbf{u}_1, \mathbf{u}_2, ..., \mathbf{u}_n$. We say that $\{\mathbf{u}_1, \mathbf{u}_2, ..., \mathbf{u}_n\}$ are **_linearly independent_** if, for each i, $\mathbf{u}_i$ is not in

$$\text{sp}\{\mathbf{u}_1, \mathbf{u}_2, ...\mathbf{u}_{i-1}, \mathbf{u}_{i+1}, ..., \mathbf{u}_n\}.$$

If $\{\mathbf{u}_1, \mathbf{u}_2, ..., \mathbf{u}_n\}$ are linearly independent, then we say that

$$U = \text{sp}\{\mathbf{u}_1, \mathbf{u}_2, ..., \mathbf{u}_n\} \text{ is an } n\text{-dimensional subspace}$$

and that $\{\mathbf{u}_1, \mathbf{u}_2, ..., \mathbf{u}_n\}$ is a **basis** for U. In particular, we say V is **m-dimensional** if

$$V = \text{sp}\{\mathbf{u}_1, \mathbf{u}_2, ..., \mathbf{u}_m\}$$

for some collection $\{\mathbf{u}_1, \mathbf{u}_2, ..., \mathbf{u}_m\}$ of m linearly independent vectors in V. If, in addition, $\langle \mathbf{u}_i, \mathbf{u}_j \rangle = 0$, $i \neq j$, and $\langle \mathbf{u}_i, \mathbf{u}_i \rangle = 1$ for all $0 \leq i, j \leq m$, then we say that $\{\mathbf{u}_1, \mathbf{u}_2, ..., \mathbf{u}_m\}$ is an **orthonormal basis** for V.

We will need the following result which is usually proved in an abstract linear algebra course:

> **Theorem (Gramm-Schmidt Orthogonalization)** *If V is an m-dimensional vector space with n-dimensional subspace U, then V has and orthonormal basis $\{\mathbf{u}_1, \mathbf{u}_2, ..., \mathbf{u}_m\}$ such that $\{\mathbf{u}_1, \mathbf{u}_2, ..., \mathbf{u}_n\}$ is an orthonormal basis for U.*

*PROBLEM 22.2   A 3-SPHERE IN 4-SPACE

We will now explore 3-dimensional spheres in 4-space, which are possibly the shape of our physical universe. We shall consider 4-space as a 4-dimensional vector space V^4.

Note that every plane in 3-space has exactly one line perpendicular to it at every point. A line is **perpendicular** to a plane if it intersects the plane and is perpendicular to every line in the plane that passes through the intersection point. In V^4 we can similarly

> **a.** *Show that every 2-dimensional subspace (a plane containing the origin $\mathbf{O}$), Π, in V^4 has an **orthogonal complement**, $\Pi^\perp$, which is a 2-dimensional subspace (plane) that intersects Π only at $\mathbf{O}$ such that every line through $\mathbf{O}$ in Π is perpendicular to every line through $\mathbf{O}$ in $\Pi^\perp$.*

This was probably proved in your linear algebra course. The easiest proof (we think) is to pick an orthonormal basis for V^4 so that, in the new coordinates, Π is the span of the first two basis vectors. This is possible by the theorem above.

> **DEFINITIONS:** Let a 3-sphere, S^3, be the collection of points in V^4 that are at a fixed distance r from $\mathbf{O}$, the origin of V^4 and the **center** of the 3-sphere. The number r is called the **radius** of

the sphere.

We define a ***great circle*** on S^3 to be the intersection of S^3 with 2-dimensional subspace (a plane through the center **O**) in V^4.

We define a ***great 2-sphere*** on S^3 to be the intersection of S^3 with any 3-dimensional subspace of V^4 (that passes through **O**).

b. *Show that every great 2-sphere in the 3-sphere has reflection-in-itself symmetry.*

Choose an orthonormal basis for V^4 so that the great 2-sphere is in the 3-subspace spanned by the first three basis elements.

c. *Show that every great circle has the symmetries in S^3 of rotation through any angle and reflection through any great 2-sphere perpendicular to the great circle. Because these are principle symmetries of a straight line in 3-space, it makes sense to call these great circles **geodesics in S^3**.*

Choose an orthonormal basis for V^4 so that the great circle is in the plane spanned by the first two basis elements.

d. *If two great circles in S^3 intersect, then they lie in the same great 2-sphere.*

Suggestions for Problem 22.2

Thinking in four dimensions may be a foreign concept to you, but believe it or not, it is possible to visualize a 4-dimensional space. Remember, the fourth dimension here is not time, but a fourth spatial dimension. We know that any two intersecting lines that are linearly independent (that do not coincide) determine a 2-dimensional plane. If we then add another line that is not in this plane, the three lines span a 3-space. When lines such as these are used as coordinate axes for a coordinate system, then they are typically taken to be orthogonal — each line is perpendicular to the others. Now to get 4-space, imagine a fourth line that is perpendicular to each of these original three. This creates the fourth dimension that we are considering.

Although we cannot experience all four dimensions at once, we can easily imagine any three at a time, and we can easily draw a picture of any two. This is the secret to looking at four dimensions. These 3- or 2-dimensional subspaces look exactly the same as any other 3-space or plane that you have seen before. This holds true for any subspace of V^4 — because, in any orthonormal basis, the basis vectors are orthogonal, any set of three of these will look the same and will determine (span) a subspace geometrically identical to our familiar 3-space, and any set of two basis elements will look like any other and will determine (span) a 2-dimensional plane.

For all of these problems, you should not be looking at projections of the 3-sphere into a plane or a 3-space, but rather looking at the part of the 3-sphere that lies in a subspace. For example, because the 3-sphere is defined as the set of points a distance r from the origin O in V^4, if you take any 3-dimensional subspace (through O) of V^4, then the part of the 3-sphere that lies in this 3-dimensional subspace is the set of points a distance r from its center O in the 3-subspace. So any 3-dimensional subspace of V^4 intersects the 3-sphere in a 2-sphere, which you know all about by now, and you can easily visualize.

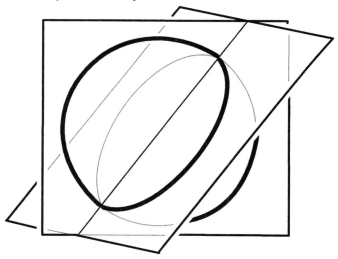

Figure 22.4   Intersecting great circles

For all of the problems here, it is generally best to draw pictures of various planes (2-dimensional subspaces) through the 3-sphere because they are easy to draw on a piece of paper. Remember, only include in your picture those geometric objects that lie in the plane you are drawing. So a great circle that lies in this plane would be drawn as a circle, while another great circle that passed through this plane would intersect this plane only in two points. See Figure 22.4.

For this particular problem, you are looking at the 3-sphere extrinsically. A good way to proceed is to draw several planes as outlined above and try to get an idea of how the planes relate to one another when combined into a 4-dimensional space. Once you have an understanding of how the different planes interact in four dimensions, it is fairly easy to show how the great circles of a 3-sphere behave.

*PROBLEM 22.3 HYPERBOLIC 3-SPACE, UPPER HALF-SPACE

As mentioned previously, there is no smooth isometric embedding of a hyperbolic plane into 3-space and, thus, no analytic isometric description. In the same way there is no isometric analytic description of a hyperbolic 3-space in 4-space. Instead we will describe hyperbolic 3-space in terms of the upper-half-space model that is analogous to the upper-half-plane model for the hyperbolic plane, which was described in Chapter 17.

> **DEFINITION:** Let $\mathbf{R}^{3+}$ = {(x, y, z) in $\mathbf{R}^3$ | $z > 0$} and call it *the upper half-space*.

In Chapter 17 we started with the annular hyperbolic plane and then defined a coordinate map z: $\mathbf{R}^{2+} \to \mathbf{H}^2$. Now we do not have an isometric model $\mathbf{H}^3$ but, instead, we have to start with the upper half-space and use z to define $\mathbf{H}^3$. Recall that z: $\mathbf{R}^{2+} \to \mathbf{H}^2$ has distortion ρ/b at the point (a,b) in $\mathbf{R}^{2+}$, where ρ is the radius of the annuli. As we saw in Chapter 17, we can study the geometry of the hyperbolic plane $\mathbf{H}^2$ by considering it to be the upper half-plane with angles as they are in $\mathbf{R}^{2+}$ and distances distorted in $\mathbf{R}^{2+}$ by ρ/b at the point (a, b). So now we use this idea to define hyperbolic 3-space $\mathbf{H}^3$.

> **DEFINITION:** Define *the upper-half-space model* of hyperbolic space $\mathbf{H}^3$ to be the upper half-space $\mathbf{R}^{3+}$ with angles as they are in $\mathbf{R}^{3+}$ and with distances distorted by ρ/c at the point (a, b, c). We call ρ the *radius* of H^3.
>
> We define a *great semicircle* in $\mathbf{H}^3$ to be the intersection of $\mathbf{H}^3$ with any circle that is in a plane perpendicular to the boundary of $\mathbf{R}^{3+}$ and whose center is in the boundary of $\mathbf{R}^{3+}$ or the

intersection of $\mathbf{R}^{3+}$ with any line perpendicular to the boundary of $\mathbf{R}^{3+}$. The ***boundary of* $\mathbf{R}^{3+}$** are those points in $\mathbf{R}^3$ with $z = 0$.

We define a ***great hemisphere*** in $\mathbf{R}^{3+}$ to be the intersection of $\mathbf{R}^{3+}$ with a sphere whose center is on the boundary of $\mathbf{R}^{3+}$ in $\mathbf{R}^3$ or the intersection of $\mathbf{R}^{3+}$ with any plane that is perpendicular to the boundary of $\mathbf{R}^{3+}$ in $\mathbf{R}^3$. See Figure 22.5.

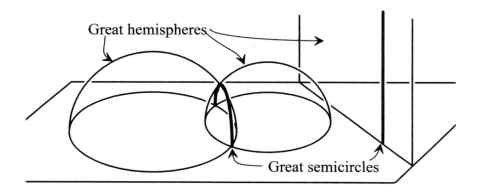

Figure 22.5 Upper-half-space model of H^3

   a. *Show that inversion through a great hemisphere in $\mathbf{R}^{3+}$ has distortion 1 in $\mathbf{H}^3$ and, thus, is an isometry in $\mathbf{H}^3$ and can be called a (hyperbolic) reflection through the great hemisphere.*

Look back at Problem **17.3**. Note that any inversion in a sphere when restricted to a plane containing the center of the sphere is an inversion of the plane in the circle formed by the intersection of the plane and the sphere.

   b. *Show that, given a great semicircle [or great hemisphere], there is a hyperbolic reflection (inversion through a great hemisphere) that takes the great semicircle [hemisphere] to a vertical half-line [half-plane] in the upper half-space.*

Look back at Problems **16.2b** and **17.3b**. Note that any inversion in a sphere when restricted to a plane containing the center of the sphere is an inversion of the plane in the circle formed by the intersection of the plane and the sphere.

Any vertical half-plane is precisely an upper-half-plane model of $\mathbf{H}^2$. Thus we conclude that each great hemisphere in $\mathbf{H}^3$ has the geometry of $\mathbf{H}^2$.

*c. *Show that every great semicircle Γ has the symmetries in $\mathbf{H}^3$ of reflection through any great hemisphere perpendicular to Γ and rotation about Γ through any angle. Because these are principal symmetries of a straight line in 3-space it makes sense to call these great semicircles **geodesics in $\mathbf{H}^3$**.*

For the reflection, look in the half-plane containing Γ and in this plane use the arguments of Problem **16.2a**. For the rotation, look at great hemispheres containing Γ and restrict your attention to their intersections with the vertical half-plane that passes through the center of Γ and is perpendicular to Γ; then refer back to Problem **5.4a**.

d. *If two great semicircles in $\mathbf{H}^3$ intersect, then they lie in the same great hemisphere.*

Use part **b** to assume that one of the great semicircles is a vertical half-line.

*PROBLEM 22.4 DISJOINT EQUIDISTANT GREAT CIRCLES

a. *Show that there are two great circles in $\mathbf{S}^3$ such that **every** point on one is a distance of one-fourth of a great circle away from **every** point on the other and vice versa.*

Is there anything analogous to this in $\mathbf{H}^3$ or in ordinary 3-space? Why?

SUGGESTIONS

This problem is especially interesting because there is no equivalent theorem on the 2-sphere; we know that on the 2-sphere, all great circles intersect, so they can't be everywhere equidistant. The closest analogy on the 2-sphere is that a pole is everywhere equidistant from the equator. When we go up to the next dimension, this pole "expands" to a great circle such that every point on this great circle is everywhere equidistant from the equator. While this may seem mind-boggling, there are ways of seeing what is happening.

An important difference created by adding the fourth dimension lies in the orthogonal complement to a plane. In 3-space, the orthogonal complement of a plane is a line that passes through a given point. This means that for any given point on the plane (the origin is always a convenient point), there is exactly one line that is perpendicular to the plane at that point. Now what happens when you add the fourth dimension? In 4-space, the orthogonal complement to a plane is a plane. This means that every line in one plane is perpendicular to every line in the other plane. To understand how this is possible, think about how it works in 3-space and refer to Figure 22.6, where we are depicting $\mathbf{R}^4$ the 4-space with x, y, z, w coordinates. Now look at the xy-plane and the zw-plane. What do you notice? Why is every line through the center in one of these planes perpendicular to every line through the center in the other?

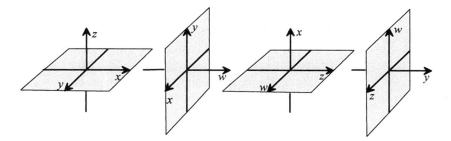

Figure 22.6   Orthogonal planes

Knowing this, look at the two great circles in terms of the planes in which they lie, and look at the relationships between these two planes, that is, where and how they intersect. Also, try to understand how great circles can be everywhere equidistant.

If we rotate along a great circle on a 2-sphere, all points of the sphere will move except for the two opposite poles of the great circle. If you rotate along a great circle on a 3-sphere, then the whole 3-sphere will move except for those points that are a quarter great circle away from the rotating great circle. Therefore, if you rotate along one of the two great circles you found above, the other great circle will be left fixed. But now rotate the 3-sphere simultaneously along both great circles at the same speed. Now every point is moved and is moved along a great circle!

b. *Write an equation for this rotation* (in *x, y, z, w* coordinates) *and check that each point of the 3-sphere is moved at the same speed along some great circle. Show that all of the great circles obtained by this rotation are equidistant from each other* (in the sense that the perpendicular distance from every point on one great circle to another of the great circles is a constant).

These great circles are traditionally called ***Clifford parallels***, named after William Clifford (1845–1879, English). See [**DG:** Thurston], pp 103–04, and [**DG:** Penrose] for readable discussions of Clifford parallels.

*PROBLEM 22.5   HYPERBOLIC AND SPHERICAL SYMMETRIES

We are now ready to see that the symmetries of great circles and great 2-spheres in a 3-sphere [and great semicircles and great hemispheres in a hyperbolic 3-space] are the same as the symmetries of straight lines and (flat) planes in 3-space. If *g* is a great circle in the 3-sphere, then let $g^{\perp}$ denote the great circle (from Problem **22.4**) every point of which is π/2 from every point of *g*.

a. *Check the entries in the table* (Figure 22.7), *which gives a summary of various symmetries of lines, great circles, and great semicircles and of (flat) planes, great 2-spheres, and great hemispheres.*

> **DEFINITION:** A surface in a 3-sphere or in a hyperbolic 3-space is called ***totally geodesic*** if, for any every pair of points on the surface, there is a geodesic (with respect to $\mathbf{S}^3$ or $\mathbf{H}^3$) that joins the two points and also lies entirely in the surface.

b. *Show that a great 2-sphere in* $\mathbf{S}^3$ *(with radius r) is a totally geodesic surface and is itself a 2-sphere of the same radius r.*

c. *Show that a great hemisphere is a totally geodesic surface in* $\mathbf{H}^3$
*(with radius r) and is isometric to a hyperbolic plane with the
same radius r.*

In the upper-half-space model there is a hyperbolic reflection that takes
every great hemisphere to a plane perpendicular to the boundary. (See
Problem **22.3**.)

Symmetries of...	Reflection through...	Reflection through...	Half-turn about...	Rotation about...	Translation along...
line $l \subset \mathbf{E}^2$	l	line $\perp l$	point in l	NA	l
great circle $g \subset \mathbf{S}^2$	g	great circle $\perp g$	point/pair in g	poles of g	g
great semicircle $g \subset \mathbf{H}^2$	g	great semicircle $\perp g$	point/pair in g	NA	g
line $l \subset \mathbf{E}^3$	plane $\supset l$	plane $\perp l$	line $\perp l$ intersecting l	l	l
great circle $g \subset \mathbf{S}^3$	great sphere $\supset g$	great sphere $\perp g$	great circle $\perp g$ intersecting g	g	g
great semicircle $g \subset \mathbf{H}^3$	great hemisphere $\supset g$	great hemisphere $\perp g$	great semicircle $\perp g$ intersecting g	g	g
plane $P \subset \mathbf{E}^3$	P	plane $\perp P$	line in P	line $\perp P$	line $\subset P$
great sphere $G \subset \mathbf{S}^3$	G	great sphere $\perp G$	great circle in G	great circle $\perp G$	great circle $\subset G$
great hemisphere $G \subset \mathbf{S}^3$	G	great hemisphere $\perp G$	great semicircle in G	great semicircle $\perp G$	great semicircle $\subset G$

Figure 22.7   Symmetries in Euclidean, spherical, and hyperbolic spaces

PROBLEM 22.6   TRIANGLES IN 3-DIMENSIONAL SPACES

Show that if A, B, C are three points in $\mathbf{S}^3$ [or in $\mathbf{H}^3$] that do not all lie on the same geodesic, then there is a unique great 2-sphere [hemisphere], $\mathbf{G}^2$, containing A, B, C.

*Thus, we can define $\triangle ABC$ as the (small) triangle in $\mathbf{G}^2$ with vertices A, B, C. With this definition, **triangles in $\mathbf{S}^3$ [or in $\mathbf{H}^3$] have all the properties that we have been studying of small triangles on a sphere [or triangles in a hyperbolic plane].***

SUGGESTIONS

Think back to the suggestions in Problems **22.2** and **22.3** — they will help you here, as well. Take two of the points, A and B, and show that they lie on a unique plane through the center, **O**, of the 3-sphere [or a unique plane perpendicular to the boundary of $\mathbf{R}^{3+}$]. Then show that there is a unique (shortest) geodesic in this plane. See Figure 22.8.

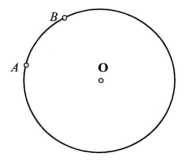

Figure 22.8   Great circle through A and B

Think of A, B, and C as defining three intersecting great circles [or semicircles]. On a 3-sphere, look at the planes in which these great circles lie and where the two planes lie in relation to one another. In hyperbolic 3-space, use a hyperbolic reflection to send one of the great semicircles to a vertical line.

Be sure that you show that the great 2-sphere (hemisphere) containing A, B, C is unique.

Chapter 23

POLYHEDRA

... if four equilateral triangles are put together, three of their plane angles meet to form a single solid angle,... When four such angles have been formed the result is the simplest solid figure ...

The second figure is composed of ... eight equilateral triangles, which yield a single solid angle from four planes. The formation of six such solid angles completes the second figure.

The third figure ... has twelve solid angles, each bounded by five equilateral triangles, and twenty faces, each of which is an equilateral triangle.

... Six squares fitted together complete eight solid angles, each composed by three plane right angles. The figure of the resulting body is the cube ...

There is still remained a fifth construction, which the god used for arranging the constellations on the whole heaven.

— Plato, *Timaeus*, 54e–55c [**AT:** Plato]

DEFINITIONS AND TERMINOLOGY

[The text in the brackets applies to polyhedra on a 3-sphere or a hyperbolic 3-space.] A ***tetrahedron***, $\triangle ABCD$, in 3-space [in a 3-sphere or a hyperbolic 3-space] is determined by any four points, A, B, C, D, called its ***vertices***, such that all four points do not lie on the same plane [great 2-sphere, great hemisphere] and no three of the points lie on the same line [geodesic]. The ***faces*** of the tetrahedron are the four [small] triangles $\triangle ABC$, $\triangle BCD$, $\triangle CDA$, $\triangle DAB$. The ***edges*** of the tetrahedron are the six line [geodesic] segments AB, AC, AD, BC, BD, CD. The ***interior*** of the tetrahedron is the [smallest] 3-dimensional region that it bounds.

335

Tetrahedra are to three dimensions as triangles are to two dimensions. Every polyhedron can be dissected into tetrahedra, but the proofs are considerably more difficult than the ones from Problem **7.5b**, and in the discussion to Problem **7.5b** there is a polyhedron that is impossible to dissect into tetrahedra without adding extra vertices. There are numerous congruence theorems for tetrahedra, analogous to the congruence theorems for triangles. We say two tetrahedra are ***congruent*** if one can be transformed by an isometry of 3-space to coincide with the other. All of the problems below apply to tetrahedra in Euclidean 3-space or a 3-sphere or a hyperbolic 3-space.

The ***dihedral angle***, $\angle AB$, at the edge AB is the angle formed at AB by $\triangle ABC$ and $\triangle ABD$. The dihedral angle is measured by intersecting it with a plane that is perpendicular to AB at a point between A and B. The ***solid angle*** at A, $\angle A$, is that portion of the interior of the tetrahedron "at" the vertex A. See Figure 23.1.

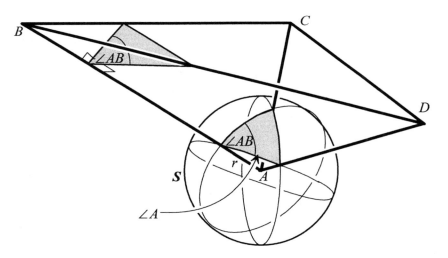

Figure 23.1   Dihedral and solid angles

You may find it helpful with these problems to construct some tetrahedra out of cardboard.

PROBLEM 23.1   MEASURE OF A SOLID ANGLE

The measure of the solid angle is defined as the ratio

$$m(\angle A) = [\lim_{r \to 0}]\text{area}\{(\text{interior of } \triangle ABCD) \cap S\}/r^2 \,,$$

where S is any small 2-sphere with center at A whose radius, r, is smaller than the distance from A to each of the other vertices and to each of the edges and faces not containing A. Note that this definition is analogous to the definition of radian measure of an angle. *Do you see why?*

    a. *Show that the measures of the solid and dihedral angles of a tetrahedron satisfy the following relationship*:

$$m(\angle A) = m(\angle AB) + m(\angle AC) + m(\angle AD) - \pi.$$

    b. *Show that two solid angles with the same measure are not necessarily congruent.* We say the two solid angles are ***congruent*** if one can be transformed by an isometry to coincide with the other.

SUGGESTIONS

Solid angles, whether in Euclidean 3-space or a 3-sphere or a hyperbolic 3-space, are closely related to spherical triangles on a small sphere around the vertex. You can think of starting with a sphere, S, and creating a solid angle by extending three sticks out from the center of the sphere. If you connect the ends of these sticks, you will have a tetrahedron. The important thing to notice is how the sticks intersect the sphere. They will obviously intersect the sphere at three points, and you can draw in the great circle arcs connecting these points. Look at the planes in which the great circles lie. In this problem you need to figure out the relationships between the angles of the spherical triangle and the dihedral angles.

    The formula given above for the definition of the solid angle uses the intersection of the interior of the solid angle with any small sphere S. This intersection is the small triangle that you just drew, and the area of the intersection is the area of the triangle. Because the measure of a solid angle is defined in terms of an area, it is possible for two solid angles to have the same measure without being congruent — they can have the same area without having the same shape.

    What you are asked to prove here is the relationship between the measure of a solid angle and the measures of its dihedral angles. Because they are closely related to spherical triangles on the small sphere, you can use everything you know about small triangles on a sphere.

PROBLEM 23.2   EDGES AND FACE ANGLES

We will study congruence theorems for tetrahedra that can be thought of as the three-dimensional analogue of triangles. A tetrahedron has 4 vertices, 4 faces, and 6 edges and we can denote it by $\triangle ABCD$, where A, B, C, D are the vertices. Figure 23.2.

Show that if $\triangle ABCD$ and $\triangle A'B'C'D'$ are two tetrahedra such that

$$\angle BAC \cong \angle B'A'C', \ \angle CAD \cong \angle C'A'D', \ \angle BAD \cong \angle B'A'D',$$
$$CA \cong C'A', \ BA \cong B'A', \ DA \cong D'A',$$

then $\triangle ABCD \cong \triangle A'B'C'D'$.

Part of your proof must be to show that the solid angles $\angle A$ and $\angle A'$ are congruent and not merely that they have the same measure.

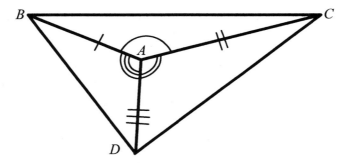

Figure 23.2   Edges and faces

SUGGESTIONS

If S is a small sphere with center at A and radius r, then

$$S \cap (\text{interior of } \triangle ABCD)$$

is a spherical triangle whose sides have lengths

$$r \ \angle BAC, \ r \ \angle CAD, \ r \ \angle BAD.$$

In the last problem, you saw how solid angles are related to spherical triangles. This problem asks you to prove the congruence of tetrahedra based on certain angle and length measurements. (Note that the angles shown above are not the dihedral angles of the tetrahedron.) So, since you can use spherical triangles to relate solid and dihedral angle measurements, why not use them to prove tetrahedra congruencies? Use the hint given to see what measurements of the spherical triangle are defined by measurements of the tetrahedron. Then see if the measurements given do in fact show congruence, and show why.

PROBLEM **23.3**   EDGES AND DIHEDRAL ANGLES

Show that if

$$AB \cong A'B', \ \angle AB \cong \angle A'B', \ AC \cong A'C',$$
$$\angle AC \cong \angle A'C', \ AD \cong A'D', \ \angle AD \cong \angle A'D',$$

then $\triangle ABCD \cong \triangle A'B'C'D'$. Figure 23.3

This is very similar to the previous problem but uses different measurements — here we have the dihedral angles instead of the angles on the faces of tetrahedron. Look at this problem the same way you looked at the previous one — see how the measurements given relate to a spherical triangle, and then prove the congruence.

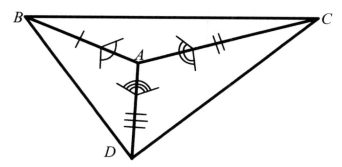

Figure 23.3   Edges and dihedral angles

PROBLEM **23.4** OTHER TETRAHEDRAL CONGRUENCE THEOREMS

Make up your own congruence theorems! Find and prove at least two other sets of conditions that will imply congruence for tetrahedra; that is, make up and prove other theorems like those in Problems **23.2** *and* **23.3**.

It is important to make sure your conditions are sufficient to prove that the solid angles are congruent, not just that they have the same measure.

PROBLEM 23.5   THE FIVE REGULAR POLYHEDRA

A *regular polygon* is a polygon lying in a plane or 2-sphere or hyperbolic plane such that all of its edges are congruent and all of its angles are congruent. For example, on the plane a regular quadrilateral is a square. On a 2-sphere and a hyperbolic plane a regular quadrilateral is constructed as shown in Figure 23.4. See also Figure 18.16 for a regular octagon on a hyperbolic plane.

Note that half of a regular quadrilateral is a Khayyam quadrilateral (see Chapter 12). On 2-spheres and hyperbolic planes there are no similar polygons; for example, a regular quadrilateral (congruent sides and congruent angles) will have the same angles as another regular quadrilateral if and only if they have the same area. (Do you see why?)

A *polyhedron* in 3-space [or in a 3-sphere or in a hyperbolic 3-space] is *regular* if all of its edges are congruent, all of its face angles are congruent, all of its dihedral angles are congruent, and all of its solid angles are congruent. The faces of a polyhedron are assumed to be polygons that lie on a plane [a great 2-sphere, a great hemisphere].

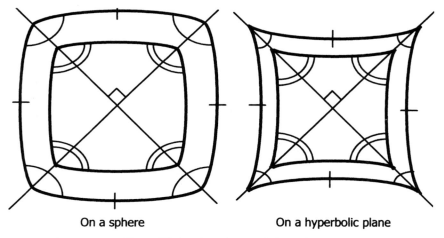

On a sphere   On a hyperbolic plane

Figure 23.4   Regular quadrilaterals

Show that there are only five regular polyhedra. In Euclidean 3-space, to say "there are only five regular polyhedra" is to mean that any regular polyhedra is similar (same shape, but not necessarily the same size) to one of the five. It still makes sense on a 3-sphere and a hyperbolic 3-space to say that "there are only five regular polyhedra," but you need to make clear what you mean by this phrase.

These polyhedra are often called the Platonic Solids, and are described by Greek philosopher Plato (429–348 B.C.) as "forms of bodies which excel in beauty" (*Timaeus*, 53e [**AT:** Plato]), but there is considerable evidence that they were known well before Plato's time. See T. L. Heath's discussion in [**AT:** Euclid, *Elements*], Vol. 3, pp. 438–39, for evidence that the five regular solids were known by Greeks before the time of Plato. In addition, there is a description of the discovery in Scotland of a complete set of the five regular polyhedra carefully carved out of stone by Neolithic persons some 4000 to 6000 years ago in [**HI:** Critchlow], pp. 148–49. The regular polyhedra are also the subject of the thirteenth (and last) book in [**AT:** Euclid, *Elements*].

SUGGESTIONS

Your argument should be essentially the same whether you are considering 3-space, or a 3-sphere, or a hyperbolic 3-space. There are many widely different ways to do this problem. The following are some approaches that we suggest:

First Approach: Note that the faces of a regular polyhedron must be regular polygons. Then focus on the vertices of regular polyhedra. Show that if the faces are regular quadrilaterals or regular pentagons, then there must be precisely three faces intersecting at each vertex. Show that it is impossible for regular hexagons to intersect at a vertex to form the solid angle of a regular polyhedron. If the faces are regular (equilateral) triangles, then show that there are three possibilities at the vertices.

Second Approach: Refer to Problem **18.5**. Each regular polyhedron can be considered to be projected out from its center onto a sphere and thus determine a cell division of the sphere. The Euler number of this spherical subdivision is $v - e + f = 2$, where v is the number of vertices, e is the number of edges, and f is the number of faces. Then

$$2e = nf \text{ and } 2e = kv . \text{ } (Why?)$$

Thus, deduce that

$$e = \frac{2}{\frac{2}{k} + \frac{2}{n} - 1},$$

and remember that e must be a positive integer.

In both approaches you should then finish the problem by using earlier problems from this chapter to show that any two polyhedra constructed from the same polygons, with the same number intersecting at each vertex, must be congruent. This step is necessary because there are polyhedra that are not rigid (that is, there are polyhedra that can be continuously moved into a non-congruent polyhedra without changing any of the faces or changing the number of faces coming together at each vertex). See Robert Connelly's "The Rigidity of Polyhedral Surfaces" (*Mathematics Magazine*, vol. 52, no. 5 (1979), pp. 275–83).

The five regular polyhedra are usually named the tetrahedron, the cube, the octahedron, the dodecahedron, and the icosahedron. (See Figure 23.5.) There is a duality (related to but not exactly the same as the duality in Chapter 20, Trigonometry and Duality) among regular polyhedra: If you pick the centers of the faces of a regular polyhedron, then these points are the vertices of a regular polyhedron, which is called the **dual** of the original polyhedron. You can see that the cube is dual to the octahedron (and vice versa), that the icosahedron is dual to the dodecahedron (and vice versa), and that the tetrahedron is dual to itself.

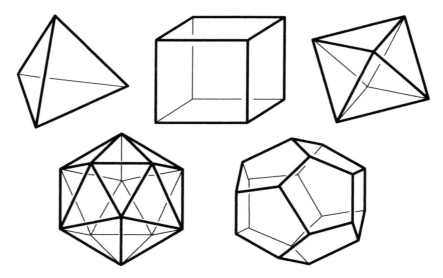

Figure 23.5   The five Platonic solids

Chapter 24

3-MANIFOLDS —
THE SHAPE OF SPACE

... if we look at the extreme points of the sky, all the visual rays appear equal to us, and if diametrically opposed stars describe a great circle, one is setting while the other is rising. If the universe, instead of being spherical, were a cone or a cylinder, or a pyramid or any other solid, it would not produce this effect on earth: one of its parts would appear larger, another smaller, and the distances from earth to heaven would appear unequal.

— Theon of Smyrna (~70–~135, Greek), [**AT:** Theon]

It will be shown that a multiply extended quantity [three-dimensional manifold] is susceptible of various metric relations, so that Space constitutes only a special case of a triply extended quantity. From this however it is a necessary consequence that the theorems of geometry cannot be deduced from general notions of quantity, but that those properties that distinguish Space from other conceivable triply extended quantities can only be inferred from experience.

— G. F. B. Riemann (1826–1866, German) *On the Hypotheses Which Lie at the Foundations of Geometry*, translated in [**DG:** Spivak], Vol. II, p. 135.

Now we come to where we live. We live in a physical 3-dimensional space, that is (at least) locally like Euclidean 3-space. The fundamental question we will investigate in this chapter is, How can we tell what the shape of our Universe is? This is the same question both Theon of Smyrna and Riemann were attempting to answer in the above quotes.

343

This is a very difficult question for which there is currently (as this is written) no clear answer. However, there are several things we can say.

SPACE AS AN ORIENTED GEOMETRIC 3-MANIFOLD

Presumably our physical Universe is globally a geometric 3-manifold. (A *geometric 3-manifold* is a space in which each point in the Universe has a neighborhood that is isometric with a neighborhood of either Euclidean 3-space, a 3-sphere, or a hyperbolic 3-space. The notion of a 3-manifold was introduced in the work by Riemann quoted above.) We say "is globally a geometric 3-manifold" in the same sense in which we say that globally Earth is a sphere (and spherical geometry is the appropriate geometry for intercontinental airplane flights) even though it is clear almost anywhere on the earth that locally there are many hills and valleys that make Earth not locally isometric to a sphere. However, the highest point on Earth (Mount Everest) is 8.85 km above sea level and the lowest point on the floor of the ocean (the Mariana Trench) is 10.99 km below sea level — the difference is only about 0.3% of the 6368 km radius of Earth (variations in the radius are of the same magnitude).

It is known that locally our physical Universe is definitely *not* a geometric 3-manifold. It was predicted by Einstein's general theory of relativity that the local curvature of our physical Universe is affected by any mass (especially large masses like our sun and other stars) [Albert Einstein (1879–1955)]. This effect is fairly accurately illustrated by imagining a 2-dimensional universe that is the surface of a flat rubber sheet. If you place steel balls on this rubber sheet, the balls will locally make dents or dimples in the sheet and thus will locally distort the flat Euclidean geometry. Einstein's prediction has been confirmed in two ways, as follows:

1. The orbits of the planets Mercury, Jupiter, and Saturn are (quite accurately) ellipses, and the major axes of these elliptical orbits change directions (precess). Classical Newtonian mechanics (based on Euclidean geometry) predicts that in a century the precession will be

    Mercury: 1.48°, Jupiter: 1.20°, Saturn: 0.77°,

    measured in degrees of an arc.

    Astronomers noticed that the observed amount of precession agreed accurately with these values for Saturn and Jupiter;

however, for Mercury (the closest planet to the Sun) the observed precession is 1.60°, which is 0.12° more than is predicted by Newtonian/Euclidean methods. But if one does the computations based on the curvature of space near the sun that is predicted by Einstein, the calculations agree accurately with the observed precession. For a mathematician's description of this calculation (which uses formulas from differential geometry), see [**DG**: Morgan], Chapter 7.

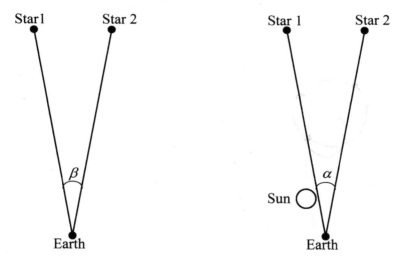

Figure 24.1   Observing local non-Euclidean geometry in the Universe

2.   In 1919 British astronomers, led by Arthur Eddington (1882–1944), measured the angle subtended by two stars from Earth: once when the Sun was not near the path of the light from the stars to Earth and once when the path of light from one of the stars went very close to the Sun. See Figure 24.1. In order to be able to see the star when its light passes close to the Sun, they had to make the second observation during a total eclipse of the Sun. They observed that the angle (α) measured with the Sun near was smaller than the angle (β) measured when the Sun was not near. The difference between the two measured angles was exactly what Einstein's theory predicted. Some accounts of this experiment talk about the Sun "bending" the light rays, but it is more accurate to say that light follows intrinsically straight paths

(geodesics) and that the Sun distorts the geometry (curvature) of the space nearby. For more details of this experiment and other experiments that verified Einstein's theory, see

http://www.ncsa.uiuc.edu/Cyberia/NumRel/EinsteinTest.html

But most of our physical Universe is empty space with only scattered planets, stars, galaxies, and the "dimples" that the stars make in space is only a very small local effect near the star. The vast empty space appears to locally (on a medium scale much larger than the scale of the distortions near stars) be a geometry of the local symmetries of Euclidean 3-space. In particular, our physical space is observed to be locally (on a medium scale) the same in all directions with isometric rotations, reflections, and translations as in Euclidean 3-space.

THEOREM 24.0. *Euclidean 3-space, 3-spheres, and hyperbolic 3-spaces are the only simply connected 3-dimensional geometries that locally have the same symmetries as Euclidean 3-space. (**Simply connected** means that every loop can be continuously shrunk to a point in the space.)*

The condition of simply connected is to rule out general geometric manifolds modeled on Euclidean 3-space, 3-spheres, or hyperbolic 3-spaces. There is a discussion of the proof (and more precise statement) of this theorem in [**DG:** Thurston], Section 3.8, where Thurston discusses eight possible 3-dimensional simply connected geometries, but only three have the same symmetries as Euclidean 3-space. A more elementary discussion without proofs can be found in [**DG:** Weeks], Chapter 18.

Unfortunately (or maybe fortunately!), geometric 3-manifolds are not fully understood. At this point no one knows what all of the geometric 3-manifolds are or how to distinguish one from the other. The theory of 3-manifolds is an area of current active research. For a very accessible discussion of this research, see [**DG:** Weeks], Part III; for more detailed discussions, see Thurston's *Three-Dimensional Geometry and Topology*, Volume 1, [**DG:** Thurston] and the second (and further?) volumes as soon as they appear.

Problem **24.1** Is Our Universe Non-Euclidean?

There is a widely reported story about the famous mathematician Carl Friedrich Gauss (1777–1855, German) that he tried to measure the angles of a triangle whose vertices were three mountain peaks in Germany. If the sum of the angles had turned out to be other than 180°, then he would have deduced that the Universe is not Euclidean (or that light does not travel in straight lines). However, his measurements were inconclusive because he measured the angles at 180° within the accuracy of his measuring instruments. This story is apparently a myth [see Breitenberger, "Gauss' Geodesy and the Axiom of Parallels," *Archive for the History of the Exact Sciences*, 31(1984), pp. 273–89], for though Gauss did measure the angles of a large triangle of mountain peaks, the purpose was to connect several grids of triangles that were being used for making a complete survey of Europe. Nevertheless, the story still leads to the question

    a. *Could we now show that the Universe is non-Euclidean by measuring the angles of a large triangle in our solar system? How accurately would we have to measure the angles?*

Note that, if the Universe is a 3-sphere or a hyperbolic 3-space, the radius R of the Universe would have to be at least as large as the diameter of our galaxy, which is about 10^{18} km. In the foreseeable future, the largest triangle whose angles we could measure has area less than the area of our solar system, which is about 8×10^{19} km^2. Use the formulas you found in Problems **7.1** and **7.2**.

    You will find that in order to determine the geometry of space, we have to look further than our solar system.

    b. *If the stars were distributed uniformly in space* (it is not clear to what extent this is actually true)*, how could you tell by looking at stars at different distances whether space was locally Euclidean, spherical, or hyperbolic?*

If you have trouble envisioning this, then start with the analogous problem for a 2-dimensional bug on a plane, sphere, or hyperbolic plane. What would this bug observe? Assume that you can tell how far away each star is — this is something that astronomers know how to do.

There are certain types of stars (called by astronomers, **standard candles**) that have a fixed known amount of brightness. These include the so-called Type Ia Supernovae.

    c. *Suppose we can see several of these standard candles and can determine accurately their distances from Earth. How could we tell from this information whether the Universe is Euclidean, spherical, or hyperbolic?*

Hint: The apparent brightness of a shining object in Euclidean space is inversely proportional to the square of the distance to the object.

    d. *It is impractical to measure the excess of triangles in our solar system by only taking measurements of angles within our solar system. What observations of only distant stars and galaxies would tell us that the Universe is not Euclidean 3-space?*

If you have trouble conceptualizing a 3-sphere or hyperbolic 3-space, then you can do this problem, first, for a very small bug on a 2-sphere or hyperbolic plane who can see distant points (stars) but who is restricted to staying inside its "solar system," which is so small that any triangle in it has excess (or defect) too small to measure. Always think intrinsically! You can assume, generally, that light will travel along geodesics, so think about looking at various objects and the relationships you would expect to find. For example, if the Universe were a 3-sphere and you could see all the way around the Universe (the distance of a great circle), how would you know that the Universe is spherical? Why? What if we could see halfway around the Universe? Or a quarter of the way around? Think of looking at stars at these distances.

All the ways discussed in parts **a**–**d** have been tried by astronomers, but, up to now, none of these observations has been accurate enough to make a definite determination.

Recently, astronomers have attempted to find at a far (but approximately known) distance in the Universe a structure whose size is known (based on physical assumptions). If such a structure is found, then it is possible to measure the angle subtended by this structure from Earth. If the geometry of the Universe is Euclidean, then the measure of the observed angle is predicted by the Law of Cosines. If the observed angle is larger than the prediction, then the Universe would be spherical and if it is smaller, then the Universe would be hyperbolic. (*Do you see why?*)

In spring 2000, a group of astronomers announced (see editorial and article in *Nature*, volume 404, April 2000) that they have observed such a structure in the cosmic background radiation and that their observations are "consistent with a flat, Euclidean Universe"; however, they also wrote that the precise details of their observations did not "agree with any known physical theory." But, even if later observations and analysis determine that the Universe *is* Euclidean, it will still leave open the question of which Euclidean 3-manifold the Universe is. We will study Euclidean 3-manifolds in the next section.

In June 2001 the United States's National Aeronautic and Space Administration (NASA) launched the Wilkinson Microwave Anisotropy Probe (WMAP). See

http://map.gsfc.nasa.gov/

for more information. WMAP has been making measurements of the cosmic background radiation since fall 2001. In 2007, the European Space Agency is scheduled to launch the Planck satellite. See

http://www.esa.int/science/planck

for more information. Plank is being designed to make even more accurate maps of the cosmic background radiation. An analysis of these maps may provide the clues we need to definitely determine the global geometry of space. To describe how this will work we must first investigate geometric 3-manifolds.

PROBLEM **24.2** EUCLIDEAN **3**-MANIFOLDS

We now consider the 3-dimensional analogue of the flat torus. Consider a cube in Euclidean 3-space with the opposite faces glued through a reflection in the plane that is midway between the opposite faces. See Figure 24.2. In this figure, we have drawn a closed straight path that starts from *A* on the bottom right edge and then hits the middle of the front face at *B*. It continues from the middle of the back face and finishes at the middle of the top left edge at a point that is glued to *A*.

 a. *Show that the cube with opposite faces glued by a reflection through the plane midway between is a Euclidean 3-manifold. That is, check that a neighborhood of each point is isometric to* *a neighborhood in Euclidean 3-space.* This Euclidean 3-manifold is call the **3-torus**.

Look separately at points (such as *A*) that are in the middle of edges, points (such as *B*) that are the middle of faces, and points that are vertices (such as *C*).

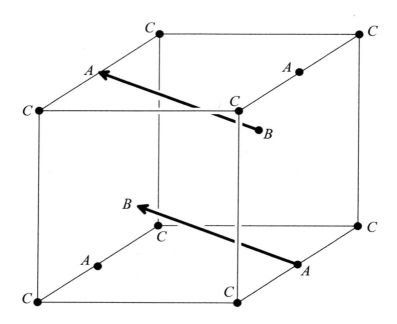

Figure 24.2   A closed geodesic path on the 3-torus

Next glue the vertical faces of the cube the same way but glue the top and bottom faces by a quarter turn. In this case we get the closed geodesic depicted in Figure 24.3.

b. *Show that you obtain a Euclidean 3-manifold from the cube with vertically opposite faces glued by a reflection through the plane midway between, and the top and bottom faces glued by a quarter-turn rotation. This Euclidean 3-manifold is called the* **quarter-turn manifold.**

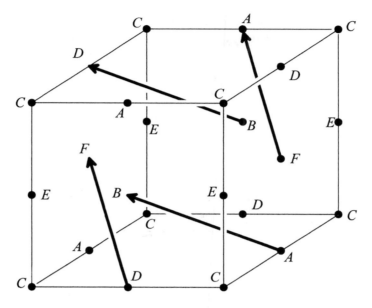

Figure 24.3   A closed geodesic in the quarter-turn manifold

c. *Draw a picture similar to Figures 24.2 and 24.3, for the **half-turn manifold**, which is the same as the quarter-turn manifold except that it is obtained by gluing the top and bottom faces with a half-turn.*

In Problem **18.1** we represented the flat torus in two different ways — one starting with a rectangle or square and the other starting from a hexagon. The above discussion of the 3-torus corresponds to the construction of the flat torus from a square. Now we want to look at what happens if we use an analogue of the hexagon construction.

Consider a hexagonal prism as in Figure 24.4. We will make gluings on the vertical sides by gluing each vertical face with its opposite in such a way that each horizontal cross-section (which are all hexagons) has the same gluings as the hexagonal flat torus (see Problem **18.2b** and Figure 18.3). The top and bottom face we glue in one of three ways. If we glue the top and bottom face through a reflection in the halfway plane, then we obtain the **hexagonal 3-torus**. If we glue the top and bottom faces with a one-sixth rotation, then we obtain the **one-sixth-turn** *manifold*. If we glue the top and bottom with a one-third rotation then we will get the ***one-third-turn manifold***.

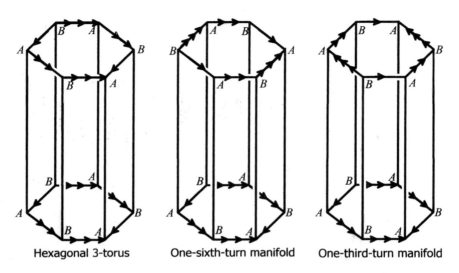

Hexagonal 3-torus   One-sixth-turn manifold   One-third-turn manifold

Figure 24.4   Hexagonal 3-manifolds

  d. *Show that the hexagonal 3-torus, the one-sixth-turn manifold, and the one-third-turn manifold are Euclidean 3-manifolds and that the hexagonal 3-torus is homeomorphic to the 3-torus. What happens if we consider the two-thirds-turn manifold and the three-sixths-turn manifold and the five-sixths-turn manifold?*

It can be shown that

THEOREM 24.2.   *There are exactly ten Euclidean 3-manifolds up to homeomorphism. Of these, four are non-orientable and six are orientable. Five of the six orientable Euclidean manifolds are the 3-torus, the quarter-turn manifold, the half-turn manifold, the one-sixth-turn manifold, and the one-third-turn manifold.*

See [**DG:** Weeks], page 252, for a discussion of this theorem. For more detail and a proof, see [**DG:** Thurston], Section 4.3.

PROBLEM 24.3 DODECAHEDRAL 3-MANIFOLDS

Spherical and hyperbolic 3-manifolds are more complicated than Euclidean 3-manifolds. In fact, no one knows what all the hyperbolic 3-manifolds are. We will only look at a few examples in order to get an idea of how to construct spherical and hyperbolic 3-manifolds in this problem and the next.

There are two examples that can be obtained by making gluings of the faces of a dodecahedron (see Problem **23.5**). It will be best for this problem for you to have a model of the dodecahedron that you can look at and touch. We want to glue the opposite faces of the dodecahedron. Looking at your dodecahedron (or Figure 24.5), you should see that the opposite faces are not lined up but rather are rotated one-tenth of a full turn from each other. Thus there are three possibilities for gluings: We can glue with a one-tenth rotation, or a three-tenths rotation, or a five-tenths (= one-half) rotation. When making the rotations it is important always to rotate in the same direction (say clockwise as you are facing the face from the outside). *You should check with your model that this is the same as rotating clockwise while facing the opposite face.*

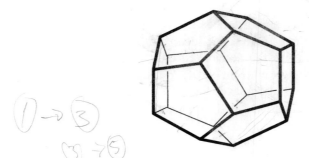

Figure 24.5   Dodecahedron

a. *When you glue the opposite faces of a dodecahedron with a one-tenth clockwise rotation, how many edges are glued together? What if you use a three-tenths rotation? Or a one-half rotation?*

The best way to do this counting is to take a model and mark one edge with tape. Then for each of the two pentagon faces on which the edge lies, the gluing glues the edge to another edge on the other side of the dodecahedron — mark those edges also. Continue with those marked edges until you have marked all the edges that are glued together.

Your answers to part **a** should be 2, 3, 5 (*not* in that order!). Thus in order for these manifolds to be geometric manifolds, we must use dodecahedrons with dihedral angles of 180°, 120°, and 72° in either Euclidean space, a sphere, or a hyperbolic 3-space. But before we go further we must figure out the size of the dihedral angles of the dodecahedron in Euclidean space, which from our model seems to be close to (if not equal to) 120°.

why?

> **b.** *Calculate the size ϕ of the dihedral angle of the (regular) dodecahedron in Euclidean 3-space.*

Imagine a small sphere with center at one of the vertices of the dodecahedron. This sphere will intersect the dodecahedron in a spherical equilateral triangle. The angles of this triangle are the dihedral angles — *Do you see why?* This triangle is called the **link of a vertex** in the dodecahedron. Determine the lengths of the sides of this triangle and then use the Law of Cosines (Problem **20.2**).

Now imagine a very small dodecahedron in 3-sphere. Its dihedral angles will be very close to the Euclidean angle ϕ (*Why is this the case?*). If you now imagine the dodecahedron growing in the 3-sphere, its dihedral angles will grow from ϕ. If you start with a very small dodecahedron in a hyperbolic space, then its dihedral angles will start very close to ϕ and then decrease as the dodecahedron grows.

> **c.** *Show that the manifold from part **a** with three edges being glued together is a spherical 3-manifold if $\phi < 120°$, or a Euclidean 3-manifold if $\phi = 120°$, or a hyperbolic 3-manifold if $\phi > 120°$.* This geometric 3-manifold is called the **Poincaré dodecahedral space** in honor of Henri Poincaré (1854–1912, French), who first described (not using the dodecahedron) a space homeomorphic to this geometric 3-manifold.

Show that each vertex of the dodecahedron is glued to three other vertices and that the four solid angles fit together to form a complete solid angle in the model (either Euclidean 3-space, 3-sphere, or hyperbolic 3-space).

d. *Can the dihedral angles of a dodecahedron in a 3-sphere grow enough to be 180°? What does such a dodecahedron look like? Is the manifold with two edges being glued together a spherical 3-manifold?* This spherical 3-manifold is called the **projective 3-space** or **RP^3**.

e. *Can the dihedral angles of a dodecahedron in a hyperbolic 3-space shrink enough to be equal to 72°? If so, the dodecahedral manifold with five edges being glued together is a hyperbolic 3-manifold.* This hyperbolic 3-manifold is called the **Seifert-Weber dodecahedral space**, after H. Seifert (1907–1996) and C. Weber, who first described both dodecahedral spaces in a 1933 article, H. Seifert and C. Weber, "Die beiden Dodekaederräume," *Mathematische Zeitschrift*, vol. 37 (1933), no. 2, p. 237.

Imagine that the dodecahedron grows until its vertices are at infinity (thus on the bounding plane in the upper half space model). Use the fact that angles are preserved in the upper-half-space model and look at the three great hemispheres that are determined by the three faces coming together at a vertex. Remember also to check that the solid angles at the vertices of the dodecahedron fit together to form a complete solid angle.

Two articles in the October 9, 2003, issue of *Nature* (vol. 425, pp. 566–567 and 593–595) showed that the spherical dodecahedral space is compatible with the latest observations of the cosmic background radiation and that it explains some measurements better than other know explanations. Note what Plato says about the dodecahedron in the quote at the beginning of Chapter 23. We will record on the Experiencing Geometry Web site the latest information as we are aware of it.

PROBLEM **24.4**   SOME OTHER GEOMETRIC 3-MANIFOLDS

We now look at three more examples of geometric 3-manifolds.

a. *Start with a tetrahedron and glue the faces as indicated in Figure 24.6. Does this gluing produce a manifold? Can the tetrahedron be put in a 3-sphere or hyperbolic 3-space so that the gluings produce a geometric 3-manifold?* We call this the **tetrahedral space**.

Figure 24.6   Tetrahedral space

b. *Start with a cube and glue each face to the opposite face with a one-quarter-turn rotation. Does this gluing produce a manifold? Can the cube be put in a 3-sphere or hyperbolic 3-space so that the gluings produce a geometric 3-manifold?* This is called the **quaternionic manifold** because its symmetries can be expressed in the quaternions (a four-dimensional version of the complex numbers with three imaginary axes and one real axis).

Again, investigate how many edges are glued together and what happens near the vertices.

c. *Start with an octahedron and glue each face to the opposite face with a one-sixth-turn rotation. Does this gluing produce a manifold? Can the octahedron be put in a 3-sphere or hyperbolic 3-space so that the gluings produce a geometric 3-manifold?* This is called the **octahedral space.**

Again, investigate how many edges are glued together and what happens near the vertices. You can also use your knowledge of solid angles from Chapter 23.

COSMIC BACKGROUND RADIATION

Astronomers from earthbound observatories have noticed a radiation that is remarkably uniform coming to Earth from all directions of space. In 1991, the United States's Cosmic Background Explorer (COBE) mapped large portions of this radiation to a resolution of about 10 degrees of arc. COBE determined that the radiation is uniform to nearly one part in 100,000, but there are slight variations (or texture) observed. It is this texture that gives us the possibility to determine the global shape of the Universe. To understand how this determination may be possible, we must first understand from where (and when!) this background radiation came.

The generally accepted explanation of the cosmic background radiation is that in the early stages of the Universe matter was so dense that no radiation could escape (the space was filled with matter so dense that all radiation was scattered). But at a certain point in time (about 300,000 years after the big bang and about 13 billion years ago), matter in the Universe started coalescing and the density of matter decreased enough that radiation could start traveling through the Universe in all directions but was still dense enough that the radiating matter (radiating because it was hot) was fairly homogeneously distributed throughout the Universe. It is this first escaped radiation that we see when we look at the cosmic background radiation.

Remember that all radiation (light and others) travels at the speed of light ($\sim 10^{13}$ kilometers per year) and thus the cosmic radiation that reaches us today has been traveling for about 13 billion years and has traveled about 1.3×10^{22} km. Thus the cosmic background radiation gives us a picture of a sphere that was a slice of the early Universe roughly 13 billion years ago. The cosmologists call this the *last scattering surface*. It appears to us that Earth is at the center of this sphere, but this is talking in space-time. What we see is a picture of a sphere in the early universe whose center (at that time) was the point in the early Universe where 13 billion years later the Milky Way galaxy, solar system, and Earth would form. By the time the radiation from this sphere reached that point, the Earth had formed and we humans are on the earth recording the radiation.

The cosmologists say that the above discussion does not involve any assumptions about the density of matter or the presence of a "cosmological constant," both of which are hotly debated subjects among cosmologists. In addition, the only assumption made about the geometry of our physical Universe is the assumption that our physical Universe is a geometric manifold and that, even though it is expanding, the topology is constant (or at least has been constant since the Universe was big enough for the radiation to escape). In Theorem **24.5**, we see that in fact if we know the topology we also know the geometry.

Before going on with the 3-dimensional discussion, let us look at an analogous situation in two dimensions. Imagine that the 2-dimensional bug's universe is a flat torus obtained from a square with opposite side glued. We may assume that the bug is at the center of the square looking out in all directions at a textured circle from the center of that circle, the last scattering circle for the bug. If this circle has diameter larger than the side of the square, then the circle will intersect itself, as indicated in Figure 24.7.

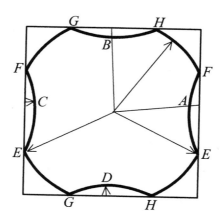

Figure 24.7   Seeing the 2-dimensional scattering circle

Note, in Figure 24.7, that when the bug in the center looks toward the point *A*, the bug will not see *A* but rather will see *C*. (The light from *A* will have reached the center of the square earlier!) Likewise, the bug will see the point *D* on the circle but not the point *B*. The important points to focus on are the points, *E*, *F*, *G*, *H*, where the circle intersects itself. The bug will see these points in two different directions. See Figure 24.7, where the point *E* is seen from both sides. If the texture is unique enough, the bug should be able to tell that it is looking at the same point in two different directions. The pattern of these identical point-pairs will indicate to the bug that its Universe is the flat torus. (See Problem **24.5a**.)

The three-dimensional situation is similar. Consider that our physical Universe is a 3-torus that is the result of gluings on a cube as in Figure 24.2. If Earth is considered to be in the center of this cube and if the sphere of last scatter has reached the faces of the cube, then the intersection of the sphere of last scatter with the faces of the cube will be circles and (because of the gluings) the circle on one face will be identified (by a reflection) with the circle on the opposite face (see Figure 24.8). The pattern of circles shows the underlying cube and the gluings. In Problem **24.5b** you will explore what the pattern of circles will look like in the other geometric 3-manifolds we have discussed.

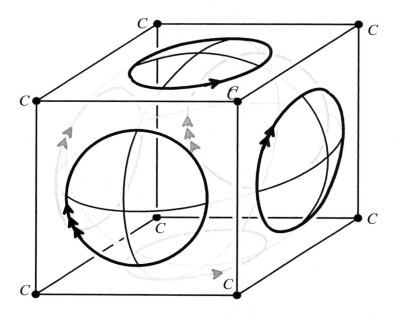

Figure 24.8   Self-intersections of the sphere of last scatter in the 3-torus

In our physical Universe, during 2002, NASA launched the Wilkinson Microwave Anisotropy Probe (WMAP). See

http://map.gsfc.nasa.gov/

for more information. In 2007, the European Space Agency is scheduled to launch the Planck satellite. See

http://astro.estec.esa.nl/

for more information. These probes will produce detailed maps of the texture of the microwave radiation. WMAP is producing about 0.3° resolution, and the Planck satellite is hoped to provide about better than 0.1° resolution. Remember that the maps before 2002 had only 10° resolution.

For further discussion of these measurements, see Luminet, Starkman, and Weeks, "Is Space Finite?", *Scientific American*, April 1999, page 90; or Cornish and Weeks, "Measuring the Shape of the Universe," *Notices of the American Mathematical Society*, 1998, or look for more recent articles.

PROBLEM 24.5 CIRCLE PATTERNS MAY SHOW THE SHAPE OF SPACE

*a. *For each of the geometric 2-manifolds in Problems* **18.1**, **18.3**, **18.4**, *what would the patterns of matching point-pairs look like if the bug's last scattering circle was large enough to intersect itself?*

Draw pictures analogous to Figure 24.7.

b. *For each of the geometric 3-manifolds in Problems* **24.2**, **24.3**, *and* **24.4**, *what would the pattern of matching circles look like if our physical Universe were the shape of that manifold **and** if the sphere of first scatter reaches far enough around the Universe for it to intersect itself?*

Draw and examine, as best you can, pictures analogous to Figure 24.8.

We saw in Problem **18.5** that the area of a spherical or hyperbolic 2-manifold is determined by its topology and the radius (or curvature) of the model. Similarly (but much more complicatedly), we have for spherical and hyperbolic 3-manifolds,

THEOREM 24.5. *If two orientable spherical or hyperbolic 3-manifolds are homeomorphic, then they are geometrically similar. That is, two such homeomorphic manifolds are isometric if the model spherical or hyperbolic spaces have the same radius (curvature).*

Thus, if we are successful in finding a pattern of circles that determines that the Universe is a spherical or hyperbolic 3-manifold, then we will know the volume of our physical Universe.

LATEST EVIDENCE ON THE SHAPE OF SPACE

We will post updates of these probes and the analyses of the data that are relevant to this text at the Experiencing Geometry Web site:

http://www.math.cornell.edu/~henderson/ExpGeom

At the time this was written there was still no conclusive evidence as to what the shape of the universe actually is. It is known that the curvature of the universe is very nearly zero and thus is very nearly a Euclidean 3-manifold, but there is a distinct possibility that the universe is a spherical 3-manifold with very large radius. However, even if the universe turns out to be a Euclidean 3-manifold, that does not mean that it is Euclidean 3-space (see Problem **24.2**). It fact, there is some evidence that suggests that if one could look far enough in the direction of the constellation Virgo, one would be able to see our galaxy! See the *Cosmology News* Web site for more discussions and links:

http://www.geometrygames.org/ESoS/CosmologyNews.html

Appendix A

EUCLID'S DEFINITIONS, POSTULATES, AND COMMON NOTIONS

> At the age of eleven, I began Euclid, with my brother as my tutor. This was one of the great events of my life, as dazzling as first love. I had not imagined that there was anything so delicious in the world.
>
> — Bertrand Russell (1883),
> *Autobiography: 1872–1914*, Allen & Unwin, 1967, p. 36

The following are the definitions, postulates, and common notions listed by Euclid in the beginning of his *Elements*, Book 1. These are Heath's translations from [**AT:** Euclid, *Elements*] except that we modified them to make the wording and usage more in line with word usage today. In our modifications we used Heath's extensive notes on the translation in order not to change the meanings involved. But, remember that we can not be sure of the exact meaning intended by Euclid — any translation should be considered only as an approximation.

DEFINITIONS

1. *A **point** is that which has no parts.*

2. *A **curve** is length without width.*
   [Heath translates this as "line," but today we normally use the term "curve" in place of "line".]

363

3. *The ends of a curve are points.*
   [It is not assumed here that the curve *has* ends (for example, see Definition 15 below); but, if the line does have ends, then the ends are points.]

4. *A (**straight**) **line** is a curve that lies symmetrically with the points on itself.*
   [The commonly quoted Heath's translation says "lies *evenly* with the points," but in his notes he says "we can safely say that the sort of idea which Euclid wished to express was that of a line ... without any irregular or unsymmetrical feature distinguishing one part or side of it from another."]

5. *A **surface** is that which has length and width only.*

6. *The boundaries of a surface are curves.*

7. *A **plane** is a surface that lies symmetrically with the straight lines on itself.*
   [The comments for Definition 4 apply here as well.]

8. *An **angle** is the inclination to one another of two curves in a plane that meet on another and do not lie in a straight line.*
   [What Euclid meant by the term "inclination" is not clear to us or, apparently, to Heath.]

9. *The angle is called **rectilinear** when the two curves are straight lines.*
   [Of course, we (and Euclid in most of the *Elements*) call these simply "angles."]

10. *When a straight line intersects another straight line such that the adjacent angles are equal to one another, then the equal angles are called **right angles** and the lines are called **perpendicular straight lines**.*
    [As discussed in the last section of Chapter 4, cones give us examples of spaces where right angles (as defined here) are not always equal to 90 degrees.]

11. *An **obtuse angle** is an angle greater than a right angle.*

12. *An **acute angle** is an angle less than a right angle.*

13. *A **boundary** of anything is that which contains it.*

14. *A **figure** is that which is contained by any boundary or boundaries.*

15. *A **circle** is a plane figure contained by one curve (called the **circumference**) with a given point lying within the figure such that all the straight lines joining the given point to the circumference are equal to one another.*

16. *The given point of a circle is called the **center** of the circle.*

17. *A **diameter** of a circle is any straight line drawn through the center and with its endpoints on the circumference; the straight line bisects the circle.*

18. *A **semicircle** is a figure contained by a diameter and the part of the circumference cut off by it. The **center** of the semicircle is the same as the center of the circle.*

19. ***Polygons** are those figures whose boundaries are made of straight lines: **triangles** being those contained by three, **quadrilaterals** those contained by four, and **multilaterals** those contained by more than four straight lines.*

20. *An **equilateral triangle** is a triangle that has three equal sides, an **isosceles triangle** is a triangle that has only two of its sides equal, and a **scalene triangle** is a triangle that has all three sides unequal.*

21. *A **right triangle** is a triangle that has a right angle, an **obtuse triangle** is a triangle that has an obtuse angle, and an **acute triangle** is a triangle that has all of its angles acute.*

22. *A **square** is a quadrilateral that is equilateral (has all equal sides) and right angled (has all right angles); a **rectangle** is a quadrilateral that is right angled but not equilateral; a **rhombus** is a quadrilateral that is equilateral but not right angled; a **rhomboid** is a quadrilateral that has opposite sides and angles equal to one another but that is neither equilateral nor right angled. Let quadrilaterals other than these be called **trapezia**.*

23. *Parallel* straight lines are straight lines lying in a plane that do not meet if continued indefinitely in both directions.

POSTULATES

1. *A (unique) straight line may be drawn from any point to any other point.*

2. *Every limited straight line can be extended indefinitely to a (unique) straight line.*

3. *A circle may be drawn with any center and any distance.*

4. *All right angles are equal.*
   [Note that cones give us examples of spaces in which all right angles are not equal; see Chapter 4. Thus this postulate could be rephrased: *"There are no cone points."*]

5. *If a straight line intersecting two straight lines makes the interior angles on the same side less than two right angles, then the two lines (if extended indefinitely) will meet on that side on which the angles are less than two right angles.*
   [See Chapter 10 for more discussion of this postulate.]

COMMON NOTIONS

1. *Things that are equal to the same thing are also equal to one another.*

2. *If equals are added to equals, then the results are equal.*

3. *If equals are subtracted from equals, the remainders are equal.*

4. *Things that coincide with one another are equal to one another.*

5. *The whole is greater than any of its parts.*

CONSTRUCTIONS OF HYPERBOLIC PLANES

We will describe four different isometric constructions of hyperbolic planes (or approximations to hyperbolic planes) as surfaces in 3-space. It is very important that you actually perform at least one of these constructions. The act of constructing the surface will give you a feel for hyperbolic planes that is difficult to get any other way. Templates for all the paper constructions (and information about possible availability of crocheted hyperbolic planes) can be found on the Experiencing Geometry Web site

www.math.cornell.edu/~henderson/ExpGeom/

THE HYPERBOLIC PLANE FROM PAPER ANNULI

A paper model of the hyperbolic plane may be constructed as follows: Cut out many identical annular ("annulus" is the region between two concentric circles) strips as in Figure B.1. Attach the strips to each other by taping the inner circle of one to the outer circle of the other. It is crucial that all the annular strips have the same inner radius and the same outer radius, but the lengths of the annular strips do not matter. You can also cut an annular strip shorter or extend an annular strip by taping two strips together along their straight ends. The resulting surface is of course only an approximation of the desired surface. The actual hyperbolic plane is obtained by letting $\delta \to 0$ while holding the radius ρ fixed. Note that since the surface is constructed (as $\delta \to 0$) the same everywhere it is *homogeneous* (that is, intrinsically and geometrically, every point has a neighborhood that is isometric to a neighborhood of any other point). We will call the results of this construction the *annular hyperbolic plane.* We strongly suggest that the reader take the time to **cut out carefully several such annuli and tape them together as indicated.**

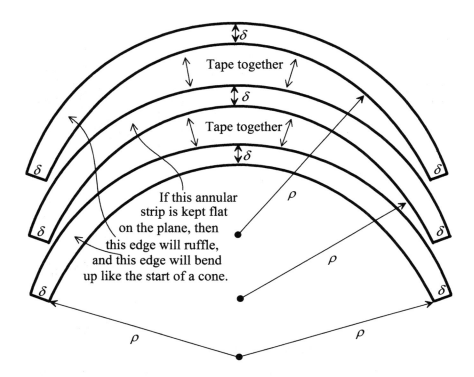

Figure B.1   Annular strips for making an annular hyperbolic plane

HOW TO CROCHET THE HYPERBOLIC PLANE

Once you have tried to make your annular hyperbolic plane from paper annuli, you will certainly realize that it will take a lot of time. Also, later you will have to play with it carefully because it is fragile and tears and creases easily — you may want just to have it sitting on your desk. But there is another way to get a sturdier model of the hyperbolic plane, which you can work and play with as much as you wish. This is the crocheted hyperbolic plane.

In order to make the crocheted hyperbolic plane, you need only basic crocheting skills. All you need to know is how to make a chain (to start) and how to single crochet. That's it! Now you can start. See Figure B.2 for a picture of these stitches, and see their description in the list below.

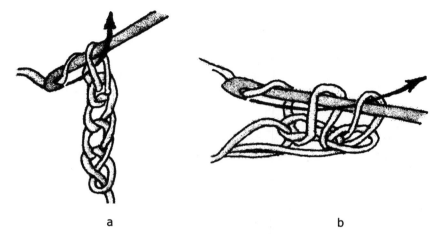

a   b

Figure B.2   Crochet stitches for the hyperbolic plane

First you should choose a yarn that will not stretch a lot. Every yarn will stretch a little but you need one that will keep its shape. Now you are ready to start the stitches:

1. Make your **beginning chain stitches** (Figure B.2a). About 20 chain stitches for the beginning will be enough.

2. **For the first stitch in each row** insert the hook into the 2nd chain from the hook. Take yarn over and pull through chain, leaving 2 loops on hook. Take yarn over and pull through both loops. One single crochet stitch has been completed (Figure B.2b).

3. **For the next N stitches** proceed exactly like the first stitch except insert the hook into the next chain (instead of the 2nd).

4. **For the $(N + 1)^{st}$ stitch** proceed as before except insert the hook into the same loop as the Nth stitch. (For example, if your $N = 12$, then crochet 12 stitches and then increase for the 13th stitch, and then 12 again and the next an increase — keep this pattern constant.)

5. **Repeat Steps 3 and 4** until you reach the end of the row.

6. **At the end of the row,** before going to the next row do one extra chain stitch.

7. **When you have the model as big as you want,** you can stop by just pulling the yarn through the last loop.

Be sure to crochet fairly tightly and evenly. That's all you need from crochet basics. Now you can go ahead and make your own hyperbolic plane. You have to increase (by the above procedure) the number of stitches from one row to the next in a constant ratio, N to $N+1$ — the ratio and size of the yarn determine the radius (the ρ in the annular hyperbolic plane) of the hyperbolic plane. You can experiment with different ratios *but* not in the same model. We suggest that you start with a ratio of 5 to 6. You will get a hyperbolic plane *only* if you will be increasing the number of stitches in the same ratio all the time.

Crocheting will take some time, but later you can work with this model without worrying about destroying it. The completed product is pictured in Figure B.3. The N for this plane is 12.

Figure B.3   A crocheted annular hyperbolic plane

{3, 7} AND {7, 3} POLYHEDRAL CONSTRUCTIONS

A polyhedral model can be constructed from equilateral triangles by putting 7 triangles together at every vertex, or by putting 3 regular heptagons (7-gons) together at every vertex. These are called the **{3, 7} polyhedral model** and the **{7, 3} polyhedral model** because triangles (3-gons) are put together **7** at a vertex, or heptagons (7-gons) are put together 3 at a vertex. These models have the advantage of being constructed more easily than the annular or crocheted models; however, one cannot make better and better approximations by decreasing the size of the triangles. This is true because at each vertex the cone angle is ($7 \times \pi/3$) = 420° or ($3 \times 5\pi/7$) = 385.71...°), no matter what the size of the triangles and heptagons are; whereas the hyperbolic plane in the small looks like the Euclidean plane with 360° cone angles. Another disadvantage of the polyhedral model is that it is not easy to describe the annuli and related coordinates.

You can make these models less "pointy" by replacing the sides of the triangles with arcs of circles in such a way that the new vertex angles are $2\pi/7$ or by replacing the sides of the heptagons with arcs of circles in such a way that the new vertex angles are $2\pi/3$. But then the model is less easy to construct because you are cutting and taping along curved edges.

See Problems **11.7** and **23.5** for more discussions of regular polyhedral tilings of plane, spheres, and hyperbolic planes.

HYPERBOLIC SOCCER BALL CONSTRUCTION

We now explore a polyhedral construction that involves two different regular polygons instead of the single polygon used in the {3, 7} and {7, 3} polyhedral constructions. A spherical soccer ball (outside the North America, called a "football") is constructed by using pentagons surrounded by five hexagons or two hexagons and one pentagon together around each vertex. The plane can be tiled by hexagons, each surrounded by six other hexagons. The hyperbolic plane can be approximately constructed by using heptagons (7-sided) surrounded by seven hexagons and two hexagons and one heptagon together around each vertex. See Figure B.4. This construction was discovered by Keith Henderson, David's son. Because a heptagon has interior angles with $5\pi/7$ radians (= 128.57...°), the vertices of this construction have cone angles of 368.57...° and thus are much smoother than the {3, 7} and {7, 3} polyhedral constructions. It also has a nice appearance if you make the heptagons a different color from the hexagons. It is also easy to construct (as long as you have a template — you can find a variety on the Experiencing Geometry Web site). As with any polyhedral construction, one cannot get closer and closer approximations to the hyperbolic plane. There is also no apparent way to see the annuli.

The hyperbolic soccer ball construction is related to the {3, 7} construction in the sense that if a neighborhood of each vertex in the {3, 7} construction is replaced by a heptagon, then the remaining portion of each triangle is a hexagon.

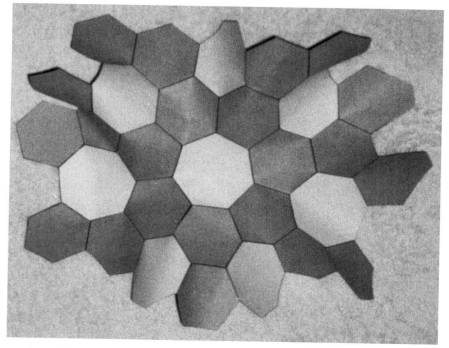

Figure B.4   The hyperbolic soccer ball constructed by Keith Henderson

"{3, 6½}" POLYHEDRAL CONSTRUCTION

We can avoid some of the disadvantages of the {3, 7} and soccer ball constructions by constructing a polyhedral annulus. In this construction we have seven triangles together only at every other vertex and six triangles together at the others. This construction still has the disadvantage of not being able to produce closer and closer approximations and it also is more "pointy" (larger cone angles) than the hyperbolic soccer ball.

The precise construction can be described in two different (but, in the end, equivalent) ways:

1. Construct polyhedral annuli as in Figure B.5 and then tape them together as with the annular hyperbolic plane.

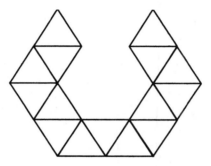

Figure B.5   Polyhedral annulus

2. The quickest way is to start with many strips as pictured in Figure B.6a — these strips can be as long as you wish. Then add four of the strips together as in Figure B.6b, using five additional triangles. Next, add another strip every place there is a vertex with five triangles and a gap (as at the marked vertices in Figure B.6b). Every time a strip is added an additional vertex with seven triangles is formed.

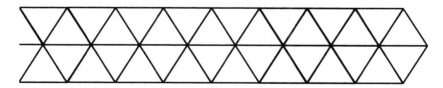

Figure B.6a   Strips

The center of each strip runs perpendicular to each annulus, and you can show that these curves (the center lines of the strip) are each geodesics because they have local reflection symmetry.

Figure B.6b   Forming the polyhedral annular hyperbolic plane

BIBLIOGRAPHY

A (cut) Short Story:
Peter wanted to know the names of the birds.
He read a book and learned the names of the birds.
Peter wanted to learn how to swim.
He read a book and drowned.
— from E. C. Basar, R. A. Bonic, et al., *Studying Freshman Calculus*, Lexington, MA: D. C. Heath and Co., 1976

This bibliography consists of books (and other items) that are referenced in this book. This is a subset of a more complete, and frequently updated, annotated bibliography with more than 500 listings on the Web at

http://www.math.cornell.edu/~henderson/biblio

Some of the references below are indicated to have online versions: For these there are clickable links in the bibliography on the Web. The section names below correspond to a subset of the section names in the bibliography on the Web.

AD ART AND DESIGN

Keith Albarn, Jenn Mial Smith, Stanford Steele, and Dinah Walker. *The Language of Pattern*. New York: Harper & Row, 1974.

William Blackwell. *Geometry in Architecture*. New York: John Wiley & Sons, 1984.

Judith Veronica Field. *The Invention of Infinity: Mathematics and Art in Renaissance*. Oxford: Oxford University Press, 1997.

375

Matila Ghyka. *The Geometry of Art and Life*. New York: Dover Publications, 1977.

Ernst Gombrich. *The Sense of Order: A Study in the Psychology of Decorative Art*. Ithaca, NY: Cornell University Press, 1978.

William M. Ivins, Jr. *Art & Geometry: A Study In Space Intuitions*. New York: Dover Publications, 1946.

Jay Kappraf. *Connections: The Geometric Bridge between Art and Science*. New York: McGraw-Hill, 1991.

AT ANCIENT TEXTS

Muhammad Ibn Musa Al'Khowarizmi. *Al-Jabr wa-l-Muqabala*. Baghdad: House of Wisdom, 825. Translation: Karpinski, L. C., editor, *Robert of Chester's Latin Translation of Al'Khowarizmi's Algebra*, New York: Macmillan, 1915.

Apollonius of Perga. *Treatise on Conic Sections*. New York: Dover, 1961.

Baudhayana. *Sulbasutram*. Bombay: Ram Swarup Sharma, 1968.

J. L. Berggren and R. S. D. Thomas. *Euclid's Phaenomena: A Translation and Study of a Hellenistic Treatise in Spherical Astronomy*. New York: Garland Publishing, 1996.

Girolamo Cardano. *The Great Art or the Rules of Algebra*. Cambridge: MIT Press, 1968.

Rene Descartes. *The Geometry of Rene Descartes*. New York: Dover Publications, Inc., 1954.

Euclid: Optics. *Journal of the Optical Society of America* 1945; 35, no. 5:357–372.

Euclid. *Euclid's Elements* (translator and editor T.L. Heath). New York: Dover, 1956.

Omar Khayyam: a paper (no title). *Scripta Mathematica* 1963; 26:323–337.

Omar Khayyam. *Risâla fî sharh mâ ashkala min musâdarât Kitâb 'Uglîdis*. Alexandria, Egypt: Al Maaref, 1958. Translated in A. R. Amir-Moez, "Discussion of Difficulties in Euclid" by Omar ibn Abrahim al-Khayyami (Omar Khayyam), *Scripta Mathematica*, 24 (1958–59), pp. 275–303.

Omar Khayyam. *Algebra*. New York: Columbia Teachers College, 1931.

Plato. *The Collected Dialogues*. Princeton, NJ: Bollinger, 1961.

Plotinus. *The Enneads* (translator: Stephen MacKenna). Burdette, NY: Larson Publications, 1992.

Proclus. *Proclus: A Commentary on the First Book of Euclid's Elements* (translator: Glenn R. Morrow). Princeton: Princeton University Press, 1970.

Theon of Smyrna. *Mathematics Useful for Understanding Plato*. San Diego: Wizards Bookshelf, 1978.

CE CARTOGRAPHY AND EARTH

L. Bagrow. *A History of Cartography*. Cabridge, MA: Harvard University Press, 1964.

Mark Monmonier. *Drawing the Line: Tales of Mapes and Cartocontroversy*. New York: Henry Holt and Company, 1995.

John P. Snyder. *Flattening the Earth: Two Thousand Years of Map Projections*. Chicago: University of Chicago Press, 1993.

Dava Sobel. *Longitude: The True Story of a Lone Genius Who Solved the Greatest Scientific Problem of His Time*. New York: Penquin Books, 1995.

DG DIFFERENTIAL GEOMETRY

C. T. J. Dodson and T. Poston. *Tensor Geometry*. London: Pitman, 1979.

David W. Henderson. *Differential Geometry: A Geometric Introduction*. Upper Saddle River, NJ: Prentice Hall, 1998.

John McCleary. *Geometry from a Differential Viewpoint*. Cambridge, UK: Cambridge University Press, 1994.

Frank Morgan. *Riemannian Geometry: A Beginner's Guide*. Boston: Jones and Bartlett, 1993.

R. Penrose. "The Geometry of the Universe." *Mathematics Today*. L. Steen (ed). New York: Springer-Verlag, 1978.

M. Santander: "The Chinese South-Seeking chariot: A simple mechanical device for visualizing curvature and parallel transport." *American Journal of Physics* 60 (9), 1992, pp. 782–787.

Michael Spivak. *A Comprehensive Introduction to Differential Geometry.* Wilmington, DE: Publish or Perish, 1979.

William Thurston. *Three-Dimensional Geometry and Topology, Vol. 1.* Princeton, NJ: Princeton University Press, 1997.

Jeffrey Weeks. *Shape of Space.* New York: Marcel Dekker, 2002.

DI DISSECTIONS

Vladimir G. Boltyanski. *Hilbert's Third Problem.* New York: John Wiley & Sons, 1978.

Vladimir G. Boltyanskii. *The Decomposition of Figures into Smaller Parts.* Chicago: University of Chicago Press, 1980.

Howard Eves. *A Survey of Geometry.* Boston: Allyn & Bacon, 1963.

Greg Frederickson. *Dissections: Plane and Fancy.* New York: Cambridge University Press, 1997.

Greg Frederickson. *Hinged Dissections: Swinging & Twisting.* Cambridge, UK: Cambridge University Press, 2002.

C.-W. Ho. "Decomposition of a Polygon into Triangles." *Mathematical Gazette* 1976; 60:132–134

Harry Lindgren. *Recreational Problems in Geometric Dissection and How to Solve Them.* New York: Dover, 1972.

Harry Lindgren. *Geometric Dissections.* Princeton, NJ: D. Van Nostrand Company, 1964.

C. H. Sah. *Hilbert's Third Problem: Scissors Congruence.* London: Pitman, 1979.

EG EXPOSITIONS — GEOMETRY

H. S. M. Coxeter and S. L. Greitzer. *Geometry Revisited.* New York: The L. W. Singer Company, 1967.

Catherine A. Gorini. *Geometry at Work: Papers in Applied Geometry.* Washington, DC: Mathematical Association of America, 2000.

David Hilbert and S. Cohn-Vossen. *Geometry and the Imagination.* New York: Chelsea Publishing Co., 1983.

Evans G. Valens. *The Number of Things: Pythagoras, Geometry and Humming Strings*. New York: E. P. Dutton and Company, 1964.

EM EXPOSITIONS — MATHEMATICS

Phillip J. Davis and Rueben Hersh. *The Mathematical Experience*. Boston: Birkhauser, 1981.

Phillip J. Davis. *The Thread: A Mathematical Yarn*. Boston: Birkhäuser Verlag, 1983.

Keith Devlin. *Mathematics: The Science of Patterns*. New York: Scientific American Library, 1994.

FO FOUNDATIONS OF GEOMETRY

David Hilbert. *Foundation of Geometry (Grundlagen der Geometrie)*. LaSalle, IL: Open Court Press, 1971.

GC GEOMETRY IN DIFFERENT CULTURES

George Bain. *Celtic Arts: The Methods of Construction*. London: Constable, 1977.

Bibhutibhushan Datta. *The Science of the Sulba*. Calcutta: University of Calcutta, 1932.

Ron Eglash. *African Fractals: Modern Computing and Indigenous Design*. New Brunswick: Rutgers University Press, 1999.

Paulus Gerdes. *Women, Art and Geometry in Southern Africa*. Trenton: Africa World Press, Inc., 1998.

John G. Neihardt. *Black Elk Speaks: Being the Life Story of a Holy Man of the Oglala Sioux*. Lincoln, NE: University of Nebraska Press, 1961.

HI HISTORY OF MATHEMATICS

J. H. Barnett. "Enter, Stage Center: the Early Drama of the Hyperbolic Functions." *Mathematics Magazine* 2004; 77:15–30.

J. L. Berggren. *Episodes in the Mathematics of Medieval Islam*. New York: Springer-Verlag, 1986.

Walter Burkert. *Lore and Science in Ancient Pythagoreanism.* Cambridge, MA: Harvard University Press, 1972.

Ronald Calinger. *A Contextual History of Mathematics: to Euler.* Upper Saddle River, NJ: Prentice Hall, 1999.

Keith Critchlow. *Time Stands Still.* London: Gordon Fraser, 1979.

David Dennis. *Historical Perspectives for the Reform of Mathematics Curriculum: Geometric Curve Drawing Devices and Their Role in the Transition to an Algebraic Description of Functions.* Ph.D. thesis. Ithaca, NY: Cornell University, 1995.

James Evans. *The History and Practice of Ancient Astronomy.* New York: Oxford University Press, 1998.

Howard Eves. *Great Moments in Mathematics (after 1650).* Dolciani Mathematical Expositions. Washington, DC: Mathematics Association of America, 1981.

Jeremy Gray. *Ideas of Space: Euclidean, Non-Euclidean and Relativistic.* Oxford: Oxford University Press, 1989.

J.L. Heilbron. *Geometry Civilized: History, Culture, and Technique.* Oxford: Clarendon Press, 2000.

George Joseph. *The Crest of the Peacock.* New York: I. B. Tauris & Sons, Ltd, 1991.

Victor J. Katz. *A History of Mathematics: An Introduction.* Reading, MA: Addison-Wesley Longman, 1998.

Morris Kline. *Mathematical Thought from Ancient to Modern Times.* Oxford: Oxford University Press, 1972.

B. A. Rosenfeld. *A History of Non-Euclidean Geometry: Evolution of the Concept of a Geometric Space.* New York: Springer-Verlag, 1989.

A. Seidenberg. "The Ritual Origin of Geometry," *Archive for the History of the Exact Sciences.* 1961; 1:488–527.

B. L. van der Waerden. *Science Awakening I: Egyptian, Babylonian, and Greek Mathematics.* Princeton Junction, NJ: The Scholar's Bookshelf, 1975.

HM HISTORY OF A MATHEMATICIAN

Elena Anne Marchisotto and James T. Smith. *The Legacy of Mario Pieri in Arithmetic and Geometry*. Boston: Birkhauser, 2004.

I. M. Yaglom. *Felix Klein and Sophus Lie: Evolution of the Idea of Symmetry in the Nineteenth Century*. Boston: Birkhäuser, 1988.

HY HYPERBOLIC GEOMETRY

N. V. Efimov. "Generation of singularities on surfaces of negative curvature" [Russian]. *Mat. Sb.* (N.S.) 1964, pp. 286–320.

Marvin J. Greenberg. *Euclidean and Non-Euclidean Geometries: Development and History*. New York: Freeman, 1980.

D. Hilbert. "Über Flächen von konstanter gaussscher Krümmung," *Transactions of the A.M.S.* 1901, pp. 87–99.

N. Kuiper. "On C^1-isometric embeddings ii." *Nederl. Akad. Wetensch. Proc.* Ser. A. 1955, pp. 683–689.

Saul Stahl. *The Poincaré Half-Plane*. Boston: Jones and Bartlett Publishers, 1993.

Wayne Zage. "The Geometry of Binocular Visual Space," *Math Magazine*, vol. 53 (5), 1980, pp. 289–294.

ME MECHANISMS

Georg Agricola. *De re metallica*. Basil: E. König, 1657. Online version available.

I. I. Artobolevskii. *Mechanisms for the Generation of Plane Curves*. New York: The Macmillan Company, 1964.

R. Connelly, E. D. Demiane, and G. Rote. "Straightening Polygonal Arcs and Convexifying Polygonal Cycles." *Discrete & Computational Geometry* 2003; 30:205–239.

L. Sprague DeCamp. *The Ancient Engineers*. New York: Ballantine Books, 1974.

George B. Dyson. *Darwin among the Machines: The Evolution of Global Intelligence*. Reading, MA: Perseus Books, 1997.

Eugene S. Ferguson. *Engineering and the Mind's Eye*. Cambridge, MA: The MIT Press, 2001.

Eugene S. Ferguson. "Kinematics of Mechanisms from the Time of Watt." *United States National Museum Bulletin*, 228, 1962, pp. 185–230.

KMODDL. *Kinematic Models for Digital Design Library*. Cornell University Libraries, 2004. http://kmoddl.library.cornell.edu

A. B. Kempe. "On a General Method of Describing Plane Curves of the n-th Degree by Linkages." *Proc. Lon. Math. Soc*; VII, 1876, pp. 213–215.

A. B. Kempe. *How to Draw a Straight Line*. London: Macmillan, 1877. Online version available.

R. S. Kirby, S. Withington, A. B. Darling, and F. G. Kilgour. *Engineering in History*. New York: McGraw-Hill Book Company, 1956.

F. C. Moon. "Franz Reuleaux: Contributions to 19th Century Kinematics and the Theory of Machines." *Applied Mechanics Reviews* 56, 2003, pp. 1–25.

Agostino Ramelli. *The Various and Ingenious Machines of Agostino Ramelli: A Classic Sixteenth-Century Illustrated Treatise on Technology*. New York: Dover Publications, 1976.

Franz Reuleaux. *The Kinematics of Machinery*. London: Macmillan and Co., 1876. Online version available.

Trevor Williams. *A History of Inventions: From Stone Axes to Silicon Chips*. New York: Checkmark Books, 2000.

Robert Yates. *Tools: A Mathematical Sketch and Model Book*. Louisiana State University, 1941.

NA GEOMETRY IN NATURE

Judith Kohl and Herbert Kohl. *The View from the Oak: The Private Worlds of Other Creatures*. New York: Sierra Club Books/Charles Scribner's Sons, 1977.

PH PHILOSOPHY OF MATHEMATICS

I. Lakatos. *Proofs and Refutations*. Cambridge: Cambridge University Press, 1976.

SG   SYMMETRY AND GROUPS

F. J. Budden. *Fascination of Groups*. Cambridge: Cambridge University Press, 1972.

C. Giacovazzo. *Fundamentals of Crystallography*. New York: Oxford University Press, 1985.

Branko Grünbaum and G. C. Shepard. *Tilings and Patterns*. New York: W. H. Freeman, 1987.

István Hargittai and Magdolna Hargittai. *Symmetry: A Unifying Concept*. Bolinas, CA: Shelter Publications, 1994.

Roger C. Lyndon. *Groups and Geometry*. New York: Cambridge University Press, 1985.

José María Montesinos. *Classical Tessellations and Three-Manifolds*. New York: Springer-Verlag, 1985.

M. Senechal. *Quasicrystals and Geometry*. Cambridge, UK: Cambridge University Press, 1995.

SP   SPHERICAL GEOMETRY

Robin Hartshorne. "Non-Euclidean III.36." *American Mathematical Monthly* 110, 2003, pp. 495–502.

Isaac Todhunter. *Spherical Trigonometry: For the Use of Colleges and Schools*. London: Macmillan, 1886. Online version available.

TP   TOPOLOGY

Donald W. Blackett. *Elementary Topology*. New York: Academic Press, 1967.

G. K. Francis and Jeffrey R. Weeks. "Conway's ZIP Proof." *American Mathematical Monthly* 106, 1999, pp. 393–399.

TX   GEOMETRY TEXTS

David A. Brannan, Matthew F. Esplen, and Jeremy J. Gray. *Geometry*. Cambridge: Cambridge University Press, 1999.

Robin Hartshorne. *Geometry: Euclid and Beyond*. New York: Springer-Verlag, 2000.

George E. Martin. *Geometric Constructions*. New York: Springer-Verlag, 1998.

Richard S. Millman and George D. Parker. *Geometry: A Metric Approach with Models*. New York: Springer-Verlag, 1981.

UN THE PHYSICAL UNIVERSE

Kitty Ferguson. *Measuring the Universe*. New York: Walker & Co, 1999.

Robert Osserman. *Poetry of the Universe: A Mathematical Exploration of the Cosmos*. New York: Anchor Books, 1995.

Index

The numbers in **bold** refer to pages on which there are definitions or statements of the indexed item.

Notes

Notes

Notes

Notes

Notes

Notes

Notes

Notes

Notes

Notes